AF704420

WITHDRAWN

THE COMPLEX FAULTING PROCESS OF EARTHQUAKES

MODERN APPROACHES IN GEOPHYSICS

VOLUME 16

Managing Editor

G. Nolet, *Department of Geological and Geophysical Sciences, Princeton University, Princeton, N.J., U.S.A.*

Editorial Advisory Board

B. L. N. Kennett, *Research School of Earth Sciences, The Australian National University, Canberra, Australia*

R. Madariaga, *Institut Physique du Globe, Université Paris VI, France*

R. Marschall, *Geco-Prakla, Prakla-Seismos GMBH, Hannover, Germany*

R. Wortel, *Department of Theoretical Geophysics, University of Utrecht, The Netherlands*

The titles published in this series are listed at the end of this volume.

THE COMPLEX FAULTING PROCESS OF EARTHQUAKES

by

JUNJI KOYAMA

Graduate School of Science, Hokkaido University, Sapporo, Japan

KLUWER ACADEMIC PUBLISHERS

DORDRECHT / BOSTON / LONDON

A C.I.P. Catalogue record for this book is available from the Library of Congress

QE
534.2
.K695
1997

ISBN 0-7923-4499-5

Published by Kluwer Academic Publishers,
P.O. Box 17, 3300 AA Dordrecht, The Netherlands.

Kluwer Academic Publishers incorporates
the publishing programmes of
D. Reidel, Martinus Nijhoff, Dr W. Junk and MTP Press.

Sold and distributed in the U.S.A. and Canada
by Kluwer Academic Publishers,
101 Philip Drive, Norwell, MA 02061, U.S.A.

In all other countries, sold and distributed
by Kluwer Academic Publishers Group,
P.O. Box 322, 3300 AH Dordrecht, The Netherlands.

Printed on acid-free paper

All Rights Reserved
© 1997 Kluwer Academic Publishers
No part of the material protected by this copyright notice may be reproduced or
utilized in any form or by any means, electronic or mechanical,
including photocopying, recording or by any information storage and
retrieval system, without written permission from the copyright owner.

Printed in the Netherlands

ACKNOWLEDGEMENTS

This is notes for a course on the Complex Faulting Process of Earthquakes delivered at Peking University, China in 1994 and 1995. The course treated the problem of complex faulting process of natural earthquakes from the view point of finite-number stochastic processes and of the scaling of complicated processes.

I would like to express my sincere gratitude to Prof. Zang Shao Xian (臧紹先), Peking University for his warm hospitality during my stimulating stays in Beijing. I am indebted to my old friend, Prof. Zheng Si Hua (鄭斯華), State Seismological Bureau, China for his continuous encouragement. I am especially obliged to Prof. Hiroaki Hara (原啓明), Tohoku University who introduced me to the world of the complex system.

Many faculty members and students of Peking University kindly assisted and supported my activities in China, to them I am very grateful. In particular, I thank Japanese faculty members and graduate and under graduate students teaching and learning at Peking University for their intimate friendship and kindness. I have learned many things from them and with them I could find solace. I have shared time with many research fellows from the United States, Germany and the United Kingdom, exchanging information and ideas about modern China and the bureaucracy and hospitality of Peking University. The time is deeply engraved on my memory with the taste of Beijing kaoya and Xihu cuyu.

I have studied the earthquake source process for many years in the seismological section of the Tohoku University. Many friends and students of the University have inspired me to this challenge and assisted me to venture into the unknown. As occasion required, some were patient enough to listen to me the whole day long, and some offered unflagging help, and others kindly prepared computer programs. For all of them I am very grateful. Dr. Alan Linde, Department of Terrestrial Magnetism, Carnegie Institution, kindly read through the whole manuscript and suggested many improvements to revise it. I owe him a great deal for his time and effort.

The Japan Society for Promotion of Science and the National Education Committee of China supported and sponsored this exchange program of study between Japan and China.

Finally, I thank my wife, son and daughter for their patience supporting my research activity and for their warm heart driving me on all the time.

PREFACE

In seismology an earthquake source is described in terms of a fault with a particular rupture size. The faulting process of large earthquakes has been investigated in the last two decades through analyses of long-period seismograms produced by advanced digital seismometry. By long-period far-field approximation, the earthquake source has been represented by physical parameters such as seismic moment, fault dimension and earthquake magnitude. Meanwhile, destruction often results from strong ground motion due to large earthquakes at short distances. Since periods of strong ground motion are far shorter than those of seismic waves at teleseismic distances, the theory of long-period source process of earthquakes cannot be applied directly to strong ground motion at short distances.

The excitation and propagation of high-frequency seismic waves are of special interest in recent earthquake seismology. In particular, the description and simulation of strong ground motion are very important not only for problems directly relevant to earthquake engineering, but also to the fracture mechanics of earthquake faulting. Understanding of earthquake sources has been developed by investigating the complexity of faulting processes for the case of large earthquakes. Laboratory results on rock failures have also advanced the understanding of faulting mechanisms. Various attempts have been made to simulate, theoretically and empirically, the propagation of short-period seismic waves in the heterogeneous real earth.

These studies revealed that the excitation of short waves is strongly dependent on the complex faulting process due to fault heterogeneities and that such waves suffer from strong attenuation and scattering effects along the propagation path. This is the dilemma we face, and it leads us to admit that the source and propagation effects are not completely distinguishable from actual observations. Therefore, it is not yet satisfactory to describe the real earthquake source in terms of heterogeneous stresses, irregular material properties, and non-uniform rupture propagations on the fault plane.

In this text, we investigate the complex faulting process of large earthquakes by considering a deterministic source (coherent rupture) and a stochastic source (incoherent ruptures) occurring simultaneously on the same fault plane. This allows us to describe the complex faulting process by physical parameters which are independent of propagation effects. Detail studies on propagation effects can be found elsewhere.

This work is intended to extend the stochastic perspective of the complex faulting process of large earthquakes. The theory provides the description of seismic radiation, acceleration, seismic energy, and earthquake magnitude due to heterogeneous faulting in a general manner. Empirical relations among these parameters have been known since soon after the dawn of instrumental seismological observation. Those relations are fully investigated to constrain the source parameters of the complex faulting process. Empirical relations among earthquake activities are also investigated. All these empirical relations are re-investigated to constrain source parameters of natural earthquakes based on the *scaling* of complex faulting process.

One earthquake occurs in a complex manner but also earthquake activity itself is very complicated. Evidence for this can be found in local earthquake swarms, regional earthquake activities, and global seismicity. New concepts of *stochastic scaling* and *non-linear scaling law* are introduced to extend the understanding of such complex phenomena. The importance of these two concepts will be discussed to understand generally the physics of complex systems in nature.

Although it seems to be expository, this study starts with a review of the classical description of the earthquake faulting process. Then, we investigate the complexity of earthquake phenomena by introducing new physical parameters.

TABLE OF CONTENTS

CHAPTER 1

CLASSICAL DESCRIPTION OF EARTHQUAKE SOURCES

1.1. EARTHQUAKE FAULT

It is widely accepted that a shallow earthquake is shear faulting on a finite fault on which the rupture spreads with a finite velocity. Mathematically, an earthquake source is represented by a displacement discontinuity vector $\vec{D}$ across the fault plane Σ. For a shear fault $\vec{D}$ is perpendicular to the normal vector $\vec{n}$ of the fault plane. The orientation of $\vec{D}$ is specified in seismology by fault strike ϕ, dip angle δ and slip angle (rake) λ. Figure 1-1 schematically illustrates the definition of these earthquake fault parameters.

Slip angle λ is usually measured counter-clockwise from the fault strike so the fault is characterized as pure strike-slip fault when λ is 0 and π. Since λ indicates the movement of the hanging wall side with respect to the footwall block, the case $\lambda = 0$ corresponds to left-lateral strike-slip faulting, where the hanging wall block moves rightward relative to the footwall block. The case $\lambda = \pi$ is for right-lateral strike-slip faulting. When $\lambda = \pm\frac{\pi}{2}$, the fault is a dip-slip fault. For $0 < \lambda < \pi$, the fault is a reverse fault, whereas $-\pi < \lambda < 0$, it is a normal fault. A thrust fault, usually found in subduction zones, is a reverse faulting with dip angle δ smaller than $\frac{\pi}{4}$.

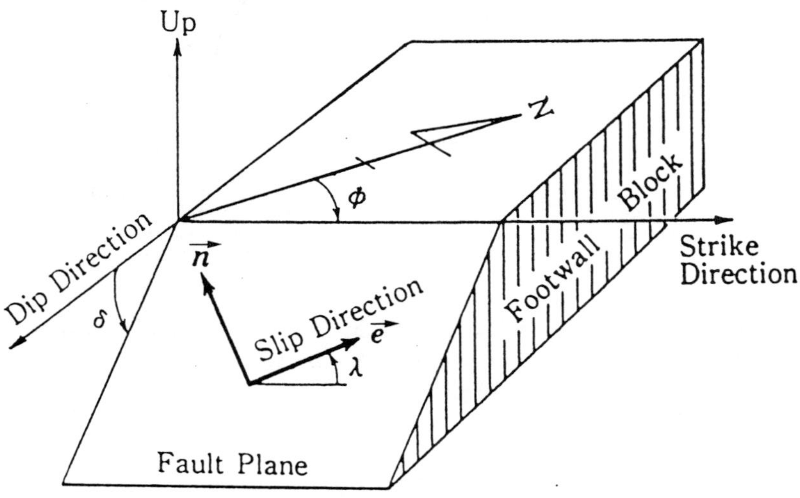

Figure 1 – 1. Schematic illustration of an earthquake fault. Displacement discontinuity on the fault is specified by fault strike ϕ, dip angle δ, and slip angle (rake) λ. Normal vector to the fault plane is $\vec{n}$ and slip direction is $\vec{e}$.

1.2. SEISMIC WAVES FROM SINGLE FORCE

Let us consider an infinite isotropic elastic medium. A Cartesian coordinate system (ξ_1, ξ_2, ξ_3) is introduced as in Fig. 1-2. The density of the medium is ρ; P and S-wave velocities are α and β, respectively. Suppose that a single force acts at the origin with a time variation $F(t)$:

$$\vec{f}(\vec{\xi}, t) = F(t)\,\vec{l}\,\delta(\xi_1)\delta(\xi_2)\delta(\xi_3), \qquad (1-1)$$

where $\vec{l}$ is the unit vector (l_1, l_2, l_3) of force $\vec{f}$ and $\delta(\xi_1)$ is the Dirac delta function. The displacement vector $\vec{u}$ at a point $\vec{x}$ has u_1, u_2, and u_3 components

$$u_i(\vec{x}, t) = \sum_{j=1}^{3} u_i^j(\vec{x}, t) l_j \quad (i = 1, 2, 3), \qquad (1-2)$$

where

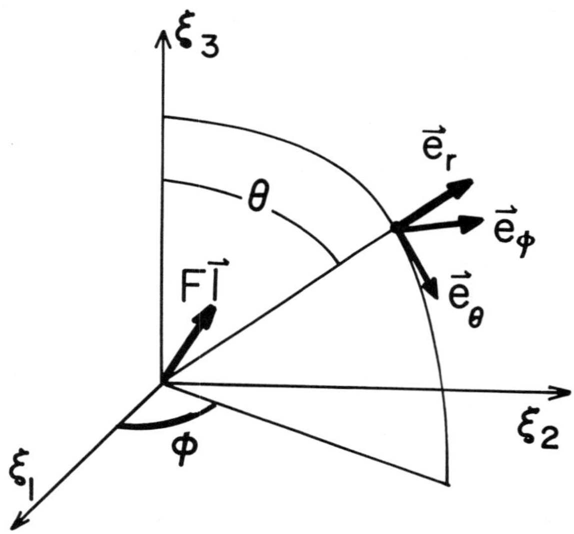

Figure 1 – 2. A Cartesian coordinate (ξ_1, ξ_2, ξ_3). Single force $F(t)\,\vec{l}$ acts at the origin in the direction of a unit vector $\vec{l}$. Unit vectors pointing radial, latitudinal, and longitudinal directions are $\vec{e}_r$, $\vec{e}_\theta$, and $\vec{e}_\phi$, respectively.

$$\begin{aligned} u_i^j(\vec{x}, t) &= \frac{3\gamma_i\gamma_j - \delta_{ij}}{4\pi\rho r_0^3} \int_{r_0/\alpha}^{r_0/\beta} \tau F(t-\tau) d\tau \\ &+ \frac{\gamma_i\gamma_j}{4\pi\rho\alpha^2 r_0} F(t - \frac{r_0}{\alpha}) + \frac{\delta_{ij} - \gamma_i\gamma_j}{4\pi\rho\beta^2 r_0} F(t - \frac{r_0}{\beta}). \qquad (1-3) \end{aligned}$$

The distance of point $\vec{x}$ from the origin is r_0, and $(\gamma_1, \gamma_2, \gamma_3)$ specifies the unit vector $\vec{\gamma}$ pointing in the direction of $\vec{x}$. δ_{ij} in (1-3) is Kronecker's delta;

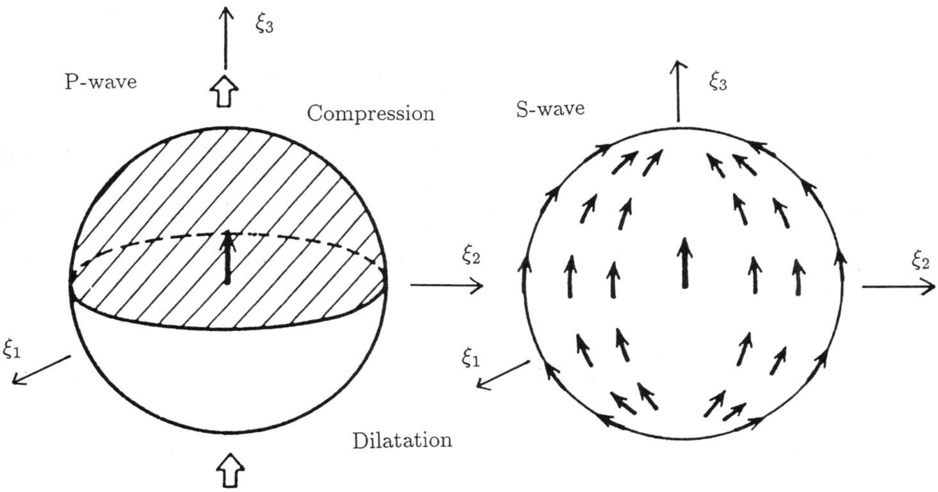

Figure 1 – 3. Radiation pattern of P and S-waves generated by a single force in ξ_3 direction. Compressive P-waves radiate from the upper hemisphere in the figure which is shaded and dilatational P-waves from the lower hemisphere. $\xi_1\xi_2$ plane is the nodal plane for P-waves. Amplitude of S-waves becomes maximum in $\xi_1\xi_2$ plane and zero in ξ_3 axis.

u_i^j represents the i-th component of displacement due to a single force in the j-th direction.

Consider a single force on the ξ_3 axis for instance. When the unit vectors in Fig. 1-2 are expressed by $\vec{e}_r$, $\vec{e}_\theta$, and $\vec{e}_\phi$, the displacement vector in the far-field due to this single force is

$$\vec{u}(\vec{x},t) = \frac{\vec{e}_r \cos\theta}{4\pi\rho\alpha^2 r_0} F(t - \frac{r_0}{\alpha}) - \frac{\vec{e}_\theta \sin\theta}{4\pi\rho\beta^2 r_0} F(t - \frac{r_0}{\beta}). \qquad (1-4)$$

The first term in the right hand side of (1-4) represents P-waves and the second S-waves. The radiation pattern for P-waves depends only on $\cos\theta$, and not on azimuth angle ϕ. Observation points within $0 < \theta < \frac{\pi}{2}$ are characterized by positive (compressional) radiation, whereas those within $\frac{\pi}{2} < \theta < \pi$ are by negative (dilatational) radiation. The radiation pattern for S-waves depends on $\sin\theta$, and it is also independent of the azimuth angle. Both radiation patterns are illustrated in Fig. 1-3.

These patterns do not match with observed radiation patterns of P and S-waves of natural earthquakes. Therefore, this force system is not an appropriate representation of the earthquake source. The radiation pattern of (1-4) may be valid only for P and S-waves generated by some special volcanic explosions.

1.3. SEISMIC WAVES FROM DOUBLE-COUPLE FORCE

A single-couple force consists of two forces, one acting at $(\frac{h}{2},0,0)$ in the positive ξ_2 direction and the other at $(\frac{-h}{2},0,0)$ in the negative ξ_2 direction. Figure 1-4 illustrates the single-couple force system. These two forces in the ξ_2 direction induce the displacement

$$u_i^2(x_1 - \frac{h}{2}, x_2, x_3, t) \; - \; u_i^2(x_1 + \frac{h}{2}, x_2, x_3, t), \tag{1 - 5}$$

where x_1, x_2, x_3 are the components of $\vec{x}$. We consider the limit of $h \to 0$

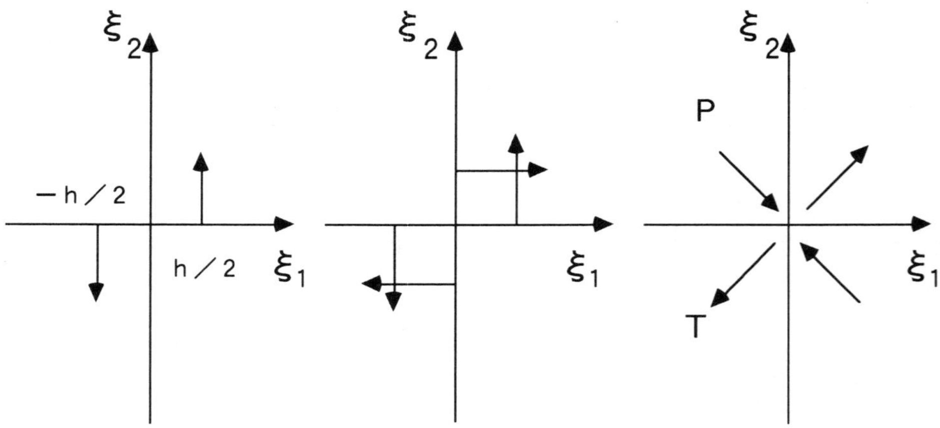

Figure 1 – 4. **Orientation of a single-couple force, a double-couple force, and the equivalent force to a double-couple point force.**

keeping the product of $F(t)h$ constant. The constant has a dimension of moment, because $F(t)$ is a force and h corresponds to an arm length:

$$M_o(t) = \lim_{h \to 0} F(t)h. \tag{1 - 6}$$

This is the single-couple point force with a seismic moment $M_o(t)$.

Obviously, from (1-5), the displacement due to this single couple is

$$u_i = -\frac{\partial u_i^2}{\partial \xi_1}, \qquad (1-7)$$

where the time function of $F(t)$ in (1-3) should be replaced by $M_o(t)$ in order to define u_i^2 in the above expression. Consequently, the displacement in the far-field is expressed by

$$u_i(\vec{x},t) = \frac{\gamma_1\gamma_2\gamma_i}{4\pi\rho\alpha^3 r_0}\dot{M}_o(t-\frac{r_0}{\alpha}) + \frac{\delta_{i2}\gamma_1 - \gamma_1\gamma_2\gamma_i}{4\pi\rho\beta^3 r_0}\dot{M}_o(t-\frac{r_0}{\beta}),$$

or

$$\begin{aligned}\vec{u}(\vec{x},t) &= \frac{\sin 2\phi \sin^2\theta}{2\pi\rho\alpha^3 r_0}\dot{M}_o(t-\frac{r_0}{\alpha})\vec{e}_r + \frac{\sin 2\phi\sin\theta\cos\theta}{2\pi\rho\beta^3 r_0}\dot{M}_o(t-\frac{r_0}{\beta})\vec{e}_\theta \\ &+ \frac{\cos^2\phi\sin\theta}{4\pi\rho\beta^3 r_0}\dot{M}_o(t-\frac{r_0}{\beta})\vec{e}_\phi. \end{aligned} \qquad (1-8)$$

where $\dot{M}_o(t)$ is the moment rate function of the single-couple force.

A double-couple force is a combination of two single-couple forces perpendicular each other as shown in Fig. 1-4. A double-couple point force acting in the $\xi_1\xi_2$ plane is

$$\begin{aligned} f_1(t) &= -M_o(t)\frac{\partial}{\partial\xi_2}\{\delta(\xi_1)\delta(\xi_2)\delta(\xi_3)\}, \\ f_2(t) &= -M_o(t)\frac{\partial}{\partial\xi_1}\{\delta(\xi_1)\delta(\xi_2)\delta(\xi_3)\}, \\ f_3(t) &= 0. \end{aligned} \qquad (1-9)$$

Since equations in (1-9) indicate the derivative of (1-1), the displacement due to (1-9) can be formally obtained as

$$u_i = -(\frac{\partial u_i^1}{\partial\xi_2} + \frac{\partial u_i^2}{\partial\xi_1}) \quad (i=1,2,3). \qquad (1-10)$$

Displacement due to a double-couple point force in the far-field is expressed by the moment rate function of $\dot{M}_o(t)$ instead of $F(t)$ in (1-3) as

$$\begin{aligned} u_i(\vec{x},t) &= \frac{2\gamma_1\gamma_2\gamma_i}{4\pi\rho\alpha^3 r_0}\dot{M}_o(t-\frac{r_0}{\alpha}) \\ &+ \frac{\delta_{i1}\gamma_2 + \delta_{i2}\gamma_1 - 2\gamma_1\gamma_2\gamma_i}{4\pi\rho\beta^3 r_0}\dot{M}_o(t-\frac{r_0}{\beta}). \end{aligned} \qquad (1-11)$$

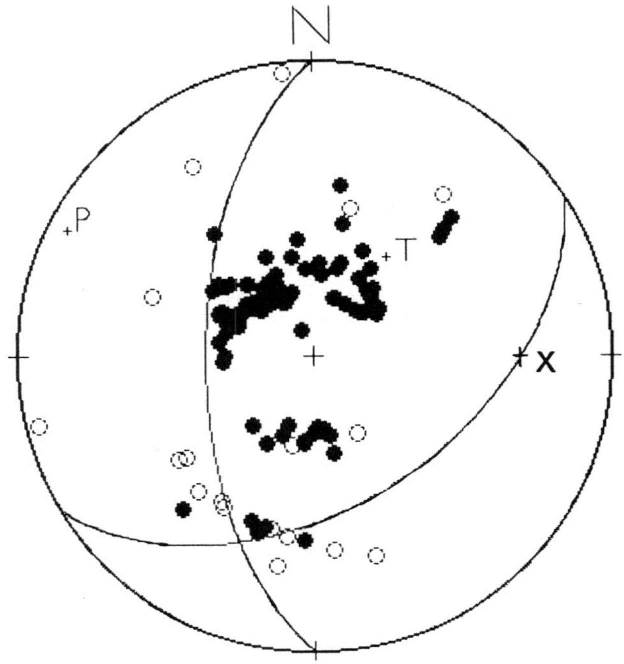

Figure 1 – 5. Initial motions of P-waves of an earthquake off the east coast of Aomori, Japan on May 16, 1978. All data are plotted on the lower focal hemisphere by an equal area projection. Open and solid circles show compression and dilatation of P-wave initial motions. One of the P-wave nodal planes is characterized by a strike of 180° and dip angle of 60° and the other by a strike of 58° and dip angle of 48°. Pressure axis marked P is characterized by azimuth angle (clock-wise from the north) of 297° and plunge angle (downward from the horizon) of 7°; tension axis T by 37° and 56°, respectively.

By using the spherical coordinates of Fig. 1-2, the displacement vector (1-11) can be rewritten as

$$\vec{u}(\vec{x},t) = \frac{\sin 2\phi \sin^2\theta}{4\pi\rho\alpha^3 r_0}\dot{M}_o(t-\frac{r_0}{\alpha})\vec{e}_r + \frac{\sin 2\phi \sin\theta\cos\theta}{4\pi\rho\beta^3 r_0}\dot{M}_o(t-\frac{r_0}{\beta})\vec{e}_\theta + \frac{\cos 2\phi \sin\theta}{4\pi\rho\beta^3 r_0}\dot{M}_o(t-\frac{r_0}{\beta})\vec{e}_\phi. \quad (1-12)$$

The first term in the right hand side of (1-12) represents P-waves and the second and third terms indicate S-waves.

Equation (1-12) shows that the polarity of P-waves from this double-couple point force depends on $\sin 2\phi$. Therefore the $\xi_1 = 0$ and $\xi_2 = 0$ planes are nodal planes for P-waves, where $\phi = 0$ or π and $\phi = \frac{\pi}{2}$ or $\frac{3\pi}{2}$, respectively. The polarity of P-waves changes on either side of these nodal planes. Therefore, the radiation pattern of P-waves shows a four-quadrant type. Almost all earthquakes show a four-quadrant type of P-wave initial motions. Figure 1-5 shows an example of P-wave initial motions and the nodal planes of P-waves for an earthquake off the east coast of Aomori, Japan on May 16, 1978.

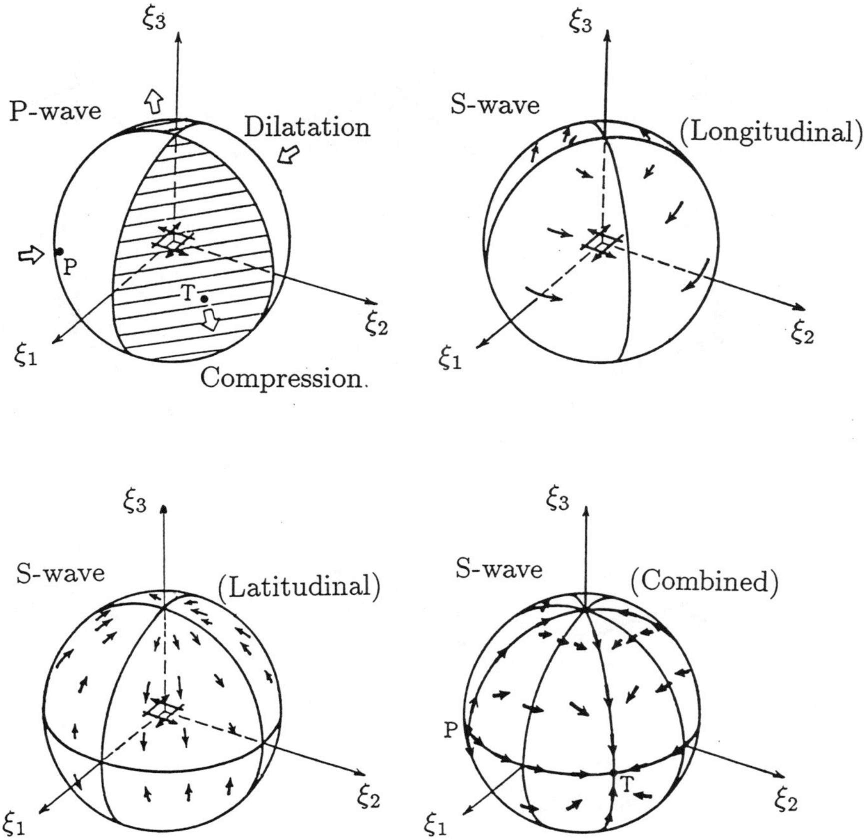

Figure 1 – 6. **Radiation pattern of P-waves, S-waves in the latitudinal direction, S-waves in the longitudinal direction, and combined S-waves. A double-couple point force is in the $\xi_1\xi_2$ plane and is illustrated by two pairs of arrows.**

Figure 1-6 illustrates the radiation patterns of P- and S-waves due to a double-couple point force on the $\xi_1\xi_2$ plane. The radiation patterns in Fig. 1-6 are very important for discriminating the focal mechanism of natural earthquakes from other force systems, and to determine the orientation of the double-couple point force. It is true that the radiation pattern of P-waves for the single couple in (1-8) is the same as that for the double couple in (1-11). The radiation pattern of S-waves for the double couple in (1-12) depends on $\sin 2\phi \sin 2\theta$ and on $\cos 2\phi \sin\theta$ for latitudinal and longitudinal components, respectively. S-wave radiation patterns also show four-quadrant types along azimuth angle ϕ. This is different from the radiation pattern for the single couple in (1-8). Consequently, S-wave observations enable us to characterize natural earthquakes as the double-couple force system.

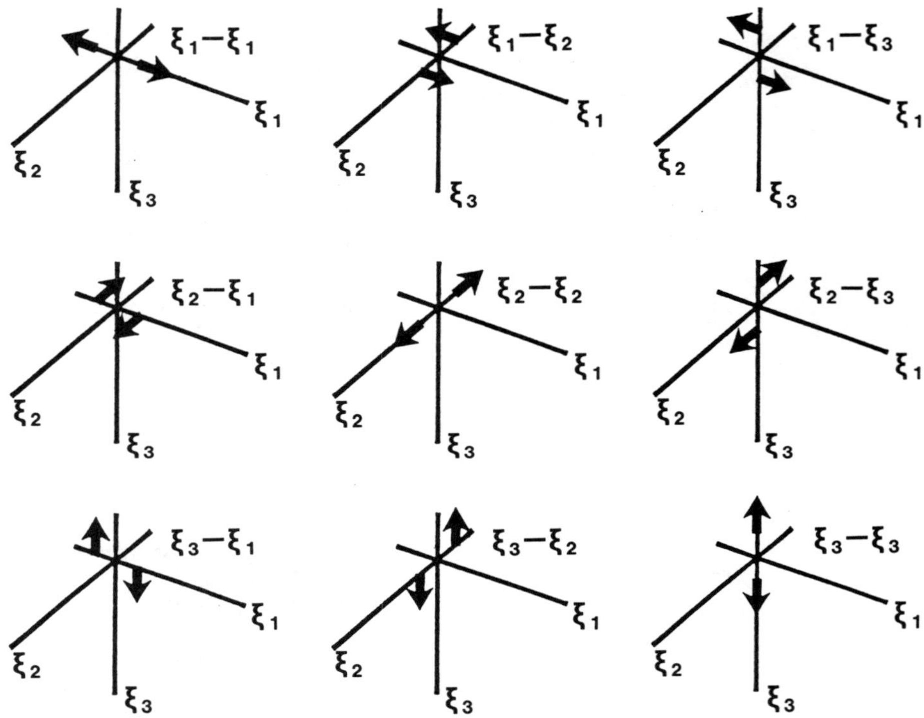

Figure 1 – 7. Possible nine couple forces to represent the general moment tensor.

1.4. MOMENT TENSOR REPRESENTATION

Generally speaking, any point source can be represented by a moment tensor M_{jk}. Similar to (1-10), displacement components due to this moment tensor can be expressed as

$$u_i(\vec{x},t) = -\Big(\frac{\partial u_i^j}{\partial \xi_k} + \frac{\partial u_i^k}{\partial \xi_j}\Big), \qquad (1-13)$$

where $M_{jk}(t)$ should be used in place of the time function $F(t)$ for the displacement u_i^j in (1-3). The derivative of a displacement component with respect to the source coordinate ξ_j is equivalent to a single couple with an arm in the ξ_j direction, as in (1-5). Thus, the moment tensor can be generally described by nine generalized couples of (ξ_1,ξ_1), (ξ_1,ξ_2), (ξ_1,ξ_3), (ξ_2,ξ_1), (ξ_2,ξ_2), (ξ_2,ξ_3), (ξ_3,ξ_1), (ξ_3,ξ_2) and (ξ_3,ξ_3). These are illustrated in Fig. 1-7. For $j = k$, the forces are in the line of the axis, these couple forces are termed vector dipoles.

1.4.1. *Moment Tensor of Double-Couple Point Force*

The moment tensor equivalent to a double couple is given by

$$
\begin{aligned}
M_{jk} &= \mu(\Delta u_j \delta\Sigma_k + \Delta u_k \delta\Sigma_j) \\
&= \mu D_0 \delta\Sigma (e_j \nu_k + e_k \nu_j), \qquad (1-14)
\end{aligned}
$$

where μ is shear modulus, Δu_j and Δu_k are the dislocation on the fault segment of $\delta\Sigma_k$ (perpendicular to ξ_k axis) and $\delta\Sigma_j$ (perpendicular to ξ_j axis), respectively. D_0 indicates an average dislocation on the fault segment $\delta\Sigma$, and $\vec{e} = (e_1, e_2, e_3)$ denotes the slip direction on the fault plane for which the normal vector is specified by $\vec{n} = (\nu_1, \nu_2, \nu_3)$. Since M_{jk} is symmetric in (1-14) in terms of j and k, the moment tensor is symmetric. In the actual application, either of the fault segments may represent the earthquake fault.

The term $e_j\nu_k + e_k\nu_j$ in (1-14) forms a tensor describing a double couple. This tensor is real and symmetric, so that, it gives real eigenvalues of $(1, 0, -1)$ and corresponding orthogonal eigenvectors. The eigenvectors specify the directions of the tension axis $\vec{t}$, the null axis $\vec{b}$, and the pressure axis $\vec{p}$ as

$$
\begin{aligned}
\vec{t} &= \frac{1}{\sqrt{2}}(\vec{n} + \vec{e}), \\
\vec{b} &= \vec{n} \times \vec{e}, \qquad (1-15) \\
\vec{p} &= \frac{1}{\sqrt{2}}(\vec{n} - \vec{e}).
\end{aligned}
$$

If fault strike ϕ, dip angle δ, and slip angle λ are as in Fig. 1-1, slip direction $\vec{e}$ and fault normal $\vec{n}$ are given by

$$
\begin{aligned}
\vec{e} &= (\cos\lambda\cos\phi + \cos\delta\sin\lambda\sin\phi)\vec{e}_x \\
&\quad + (\cos\lambda\sin\phi - \cos\delta\sin\lambda\cos\phi)\vec{e}_y \\
&\quad - \sin\delta\sin\lambda\,\vec{e}_z, \qquad (1-16)
\end{aligned}
$$

and

$$
\vec{n} = -\sin\delta\sin\phi\,\vec{e}_x + \sin\delta\cos\phi\,\vec{e}_y - \cos\delta\,\vec{e}_z, \qquad (1-17)
$$

where the geographical coordinates (x, y, z) are north, east, and downward, respectively.

Equation (1-14) together with (1-16) and (1-17) leads to the moment tensor components in geographical coordinates as

$$
\begin{aligned}
M_{xx} &= -M_o(\sin^2\phi \sin\lambda \sin 2\delta + \sin 2\phi \cos\lambda \sin\delta),\\
M_{yy} &= M_o(-\cos^2\phi \sin\lambda \sin 2\delta + \sin 2\phi \cos\lambda \sin\delta),\\
M_{zz} &= M_o(\sin 2\delta \sin\lambda),\\
M_{xy} &= M_o(\sin 2\phi \sin\lambda \sin 2\delta/2 + \cos 2\phi \cos\lambda \sin\delta),\\
M_{xz} &= -M_o(\cos\phi \cos\lambda \cos\delta + \sin\phi \sin\lambda \cos 2\delta),\\
M_{yz} &= -M_o(\sin\phi \cos\lambda \cos\delta - \cos\phi \sin\lambda \cos 2\delta), \qquad (1-18)
\end{aligned}
$$

where scalar seismic moment M_o of the fault segment $\delta\Sigma$ is

$$M_o = \mu D_0 \delta\Sigma. \qquad (1-19)$$

1.4.2. *Moment Tensor of Vertical Strike Slip Fault*

For fault strike $\phi = 0$, the dip angle $\delta = 90°$, and the slip angle $\lambda = 0$, the focal mechanism is that of a vertical strike slip fault. The slip direction on the fault is $\vec{e} = (1,0,0)$ and the vector normal to the fault plane is $\vec{n} = (0,1,0)$. The moment tensor in this case can be determined from (1-14), (1-16), and (1-17) as

$$\boldsymbol{M} = \begin{pmatrix} 0 & M_o & 0 \\ M_o & 0 & 0 \\ 0 & 0 & 0 \end{pmatrix}. \qquad (1-20)$$

The eigenvalues and eigenvectors of this tensor can be evaluated as follows: Let $\boldsymbol{M}$ be a moment tensor of second order(rank), represented as a 3×3 matrix in given coordinates. Assume that there is a vector $\vec{a}$ and scalar m such that

$$\boldsymbol{M}\vec{a} = m\vec{a},$$

where $\vec{a}$ is called the eigenvector of $\boldsymbol{M}$ and m is the corresponding eigenvalue. Solving the above equation, we transform

$$(\boldsymbol{M} - m\boldsymbol{I})\vec{a} = 0, \qquad (1-21)$$

where $\boldsymbol{I}$ is the identity matrix. Equation (1-21) is a system of three simultaneous linear equations in $\vec{a}_j$ $(j = 1,2,3)$. Non-trivial solutions are found by solving the secular equation of

$$det(\boldsymbol{M} - m\boldsymbol{I}) = 0, \qquad (1-22)$$

where *det* means the determinant. Equation (1-22) is a polynomial of third degree. It has three real roots (eigenvalues), since the moment tensor is real and symmetric. Substituting each eigenvalue m_j into (1-21), we obtain corresponding eigenvector $\vec{a}_j$.

For the moment tensor (1-20), the determinant in (1-22) reduces to

$$m^3 - {M_o}^2 m = 0, \tag{1-23}$$

giving the eigenvalues of $m = M_o$, 0, and $-M_o$. The eigenvectors in geographical coordinates can be calculated from (1-21) by applying the eigenvalues in the above, as $(\frac{1}{\sqrt{2}}, \frac{1}{\sqrt{2}}, 0)$, $(0, 0, -1)$ and $(\frac{-1}{\sqrt{2}}, \frac{1}{\sqrt{2}}, 0)$, respectively. The focal mechanism is shown in Fig. 1-8a.

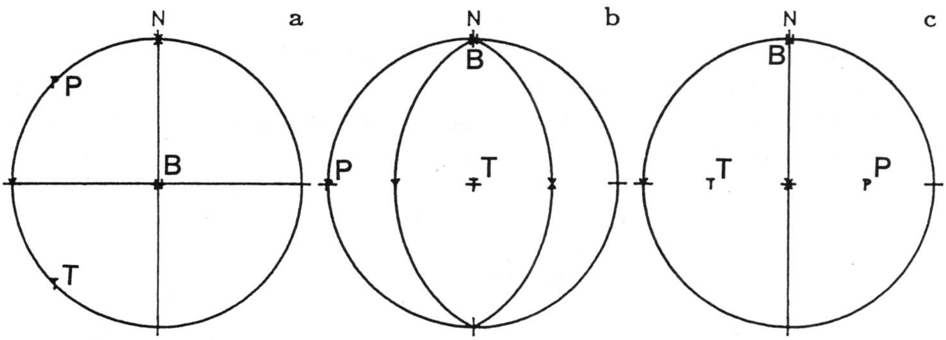

Figure 1 - 8. **Double-couple focal mechanisms. (a) A vertical strike slip fault, where fault strike is 0, dip angle 90°, and slip angle 0. (b) A 45° dip slip fault, where fault strike is 0, dip angle 45°, and slip angle 90°. (c) A vertical dip slip fault, where fault strike is 0, dip angle 90°, and slip angle 90°. The compression, tension, and null axes are indicated by P, T, and B, respectively.**

The eigenvectors are orthogonal and orthonormal. Knowing the eigenvectors, we can diagonalize $\boldsymbol{M}$ (principal axis transformation). Let $\boldsymbol{A}$ be the matrix whose columns are the orthonormal eigenvectors $\vec{a}$ of $\boldsymbol{M}$. From the orthonormality, $\boldsymbol{A}^T = \boldsymbol{A}^{-1}$, where the superscript T denotes the transpose. Then $\boldsymbol{A}^T\boldsymbol{M}\boldsymbol{A} = \vec{m}$, where $\vec{m}$ is diagonal, consisting of the eigenvalues of $\boldsymbol{M}$.

1.4.3. *Moment Tensor of 45° Dip Slip Fault*

For fault strike $\phi = 0$, the dip angle $\delta = 45°$ and the slip angle $\lambda = 90°$, the focal mechanism represents a 45° dip slip fault. The slip direction in this case is $\vec{e} = (0, \frac{-1}{\sqrt{2}}, \frac{-1}{\sqrt{2}})$ and the fault normal is $\vec{n} = (0, \frac{1}{\sqrt{2}}, \frac{-1}{\sqrt{2}})$. The moment tensor in this case is calculated from (1-14), (1-16), and (1-17) as

$$M = \begin{pmatrix} 0 & 0 & 0 \\ 0 & -M_o & 0 \\ 0 & 0 & M_o \end{pmatrix}. \qquad (1-24)$$

The corresponding eigenvalues are obtained similarly as above from (1-22) and (1-24). The determinant is the same as (1-23) and we obtain the eigenvalues of $m = M_o$, 0 and $-M_o$, and the corresponding eigenvectors are (0, 0, -1), (-1, 0, 0) and (0, 1, 0), respectively. Azimuth and plunge angles of the tension, null and pressure axes can be calculated from (1-15) as (180, 90), (360, 0) and (270, 0) in degrees. This is shown in Fig. 1-8(b), where azimuth angle is measured clockwise from the north and plunge angle is downward from the horizon.

1.4.4. *Moment Tensor of Vertical Dip Slip Fault*

A focal mechanism represents a vertical dip slip fault when $\phi = 0$, $\delta = 90°$ and $\lambda = 90°$. The slip direction in this case is $\vec{e} = (0,\ 0,\ -1)$ and the fault normal is $\vec{n} = (0,\ 1,\ 0)$. The moment tensor is calculated from (1-14), (1-16) and (1-17) as

$$M = \begin{pmatrix} 0 & 0 & 0 \\ 0 & 0 & -M_o \\ 0 & M_o & 0 \end{pmatrix}. \qquad (1-25)$$

The eigenvalues are calculated similarly as $m = M_o$, 0 and $-M_o$, and the eigenvectors are $(0, \frac{1}{\sqrt{2}}, \frac{-1}{\sqrt{2}})$, ($-1$, 0, 0) and $(0, \frac{1}{\sqrt{2}}, \frac{1}{\sqrt{2}})$. Azimuth and plunge angles of the tension, null and pressure axes can be calculated from (1-15) as (270, 45), (180, 0) and (90, 45) in degrees. This mechanism is also shown in Fig. 1-8(c).

1.5. MOMENT TENSOR OF VOLCANIC EARTHQUAKES

In the case of the 1983 eruption of the Miyakejima volcano, Japan, many of short-period volcanic earthquakes were characterized by the same direction of first motions of P-waves at all seismographic sites. Example seismograms are shown in Fig. 1-9. The initial motions of P-waves are all upward (compression) in Fig. 1-9(a), and all downward (dilatation) in Fig. 1-9(b). Although the examples in Fig. 1-9 are from a subset of observation stations, data from the extensively distributed stations exhibit the same characteristics of P-wave polarities. Figure 1-10 shows the distribution of P-wave initial motions plotted on the lower focal hemisphere. The data cover a wide area of the focal sphere with the same polarity of P-wave initial motions.

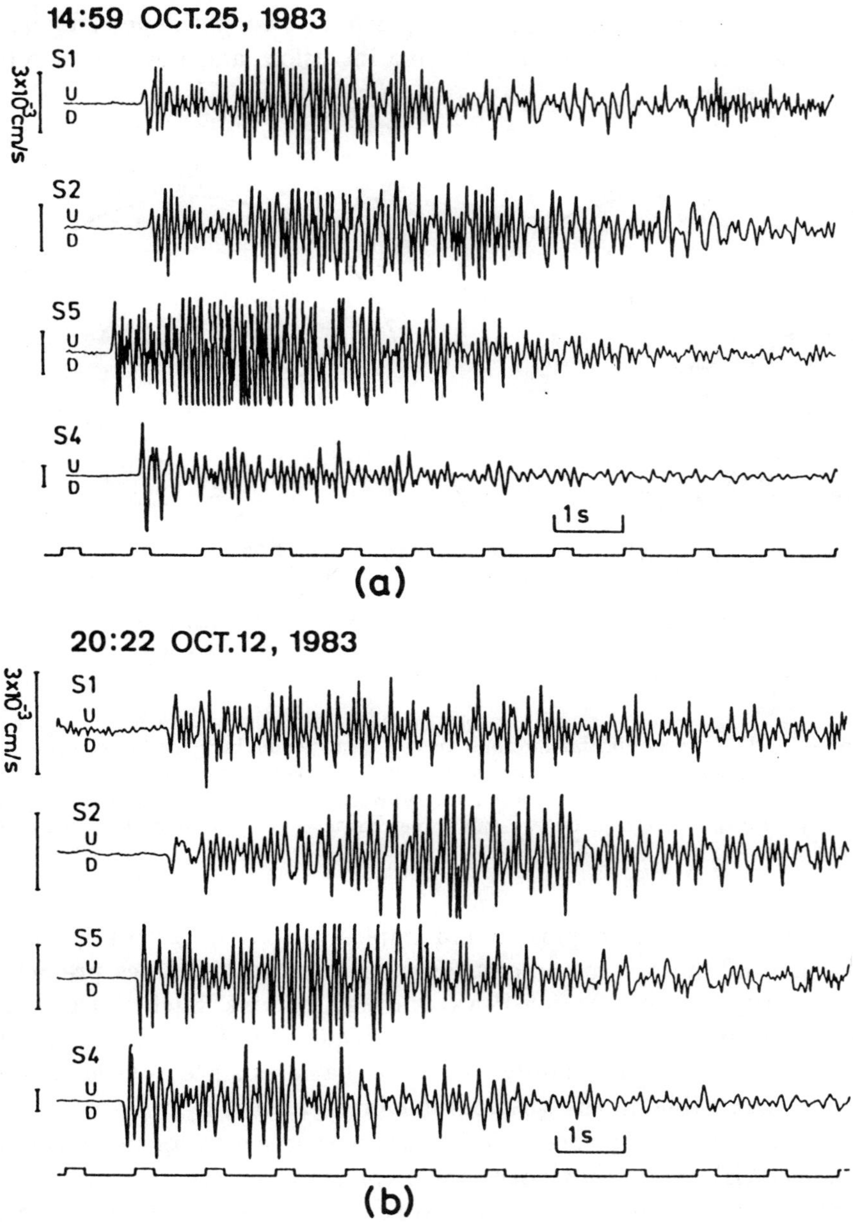

Figure 1 – 9. Short-period seismograms of volcanic earthquakes following the 1983 Miyakejima eruption, Japan (Shimizu et al., 1987). (a) P-wave initial motions are upward at all stations. (b) P-wave initial motions are downward at all stations.

These observations can not be explained by the standard quadrant-type P-wave initial motions due to the double-couple model. This is evidence that the generation of volcanic earthquakes is strongly dependent on the local stress field induced by the magma intrusion, and not due merely to tectonic stresses.

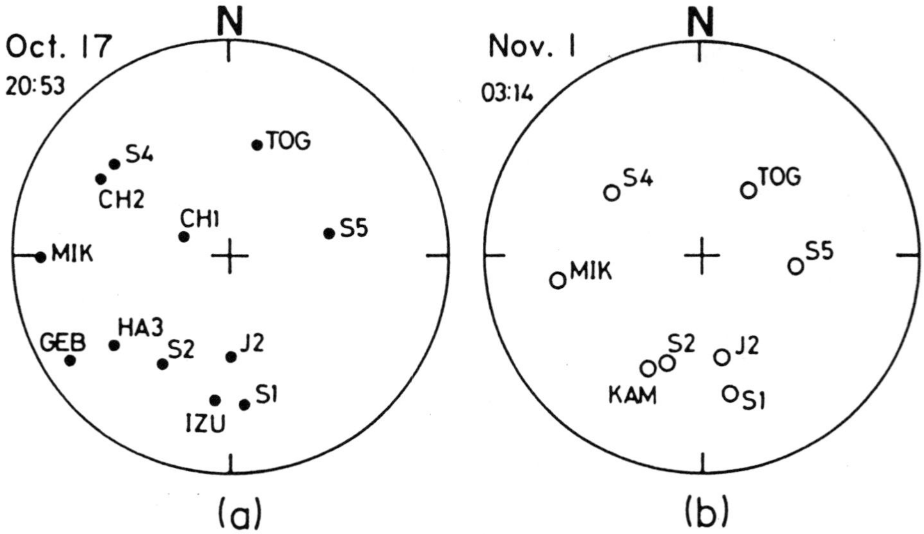

Figure 1 – 10. P-wave initial motions on the focal hemisphere for (a) compressional and (b) dilatational earthquakes following the 1983 Miyakejima eruption, Japan (Shimizu et al., 1987). Solid and open circles indicate upward and downward initial motions, respectively. Stations indicated by two and three character codes were distributed all over Miyakejima.

One may consider explosive and implosive sources as the generating mechanism of these volcanic earthquakes, since both sources predict the same polarity everywhere for P-wave initial motions. Such sources, however, would not generate S-waves since the equivalent force system is point-symmetric and thus they fail as viable mechanisms since S-wave energy is clearly recognizable on the seismograms in Fig. 1-9.

A tensile crack has been considered as the source mechanism for acoustic emissions (AE) occurring in rocks under compression tests. First motions of P-waves radiated from a tensile crack are compression in all directions. Since observed volcanic earthquakes (Fig. 1-9) have been activated by a fissure eruption on the flank of the Miyakejima volcano, the source mechanism is presumably closely related to the dikes or cracks beneath the fissure formed by the magma intrusion. Although the tensile crack radiates S-waves, the energy would not be sufficient to explain the observation.

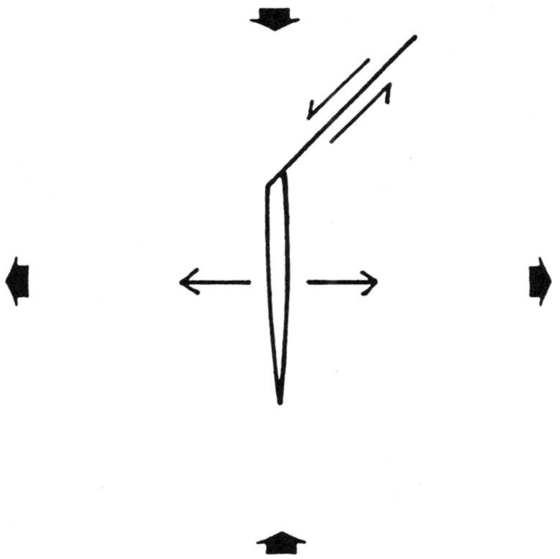

Figure 1 - 11. Schematic illustration of a tensile crack coupled with a shear crack (a tensile-shear crack model). The tension axis of the shear crack coincides with that of the tensile crack.

A tensile crack coupled with a shear crack as shown in Fig. 1-11 is considered as a candidate mechanism to explain the observations. The displacement components of P- and S-waves from the tensile-shear crack in an infinite medium are given as follows. Let $\vec{u}_P$, $\vec{u}_{SV}$, and $\vec{u}_{SH}$ be displacement vectors for P-, SV-, and SH-waves at position $\vec{x}$ and time t from a point source at $\vec{\xi}$:

$$\begin{aligned}
\vec{u}_P(\vec{x},t) &= G_P\{\vec{e}_r(\vec{\xi})^T\dot{M}(t-\frac{r_0}{\alpha})\vec{\gamma}\}\ \vec{e}_r(\vec{x}), \\
\vec{u}_{SV}(\vec{x},t) &= G_S\{\vec{e}_\theta(\vec{\xi})^T\dot{M}(t-\frac{r_0}{\beta})\vec{\gamma}\}\ \vec{e}_\theta(\vec{x}), \qquad (1-26) \\
\vec{u}_{SH}(\vec{x},t) &= G_S\{\vec{e}_\phi(\vec{\xi})^T\dot{M}(t-\frac{r_0}{\beta})\vec{\gamma}\}\ \vec{e}_\phi(\vec{x}),
\end{aligned}$$

where G_P and G_S represent the geometrical spreading for P- and S-waves. In addition, $\vec{e}_r$, $\vec{e}_\theta$, and $\vec{e}_\phi$ are unit vectors denoting the directions of particle motions for P-, SV- and SH-waves, respectively. Vector $\vec{\gamma}$ described by azimuth angle ϕ and take-off angle i, is the unit vector of a ray leaving the source point $\vec{\xi}$.

$$\begin{aligned}
\vec{e}_r(\vec{\xi}) &= \vec{\gamma} \\
&= (\sin i\cos\phi, \sin i\sin\phi, \cos i), \\
\vec{e}_\theta(\vec{\xi}) &= (\cos i\cos\phi, \cos i\sin\phi, -\sin i), \qquad (1-27) \\
\vec{e}_\phi(\vec{\xi}) &= (-\sin\phi, \cos\phi, 0).
\end{aligned}$$

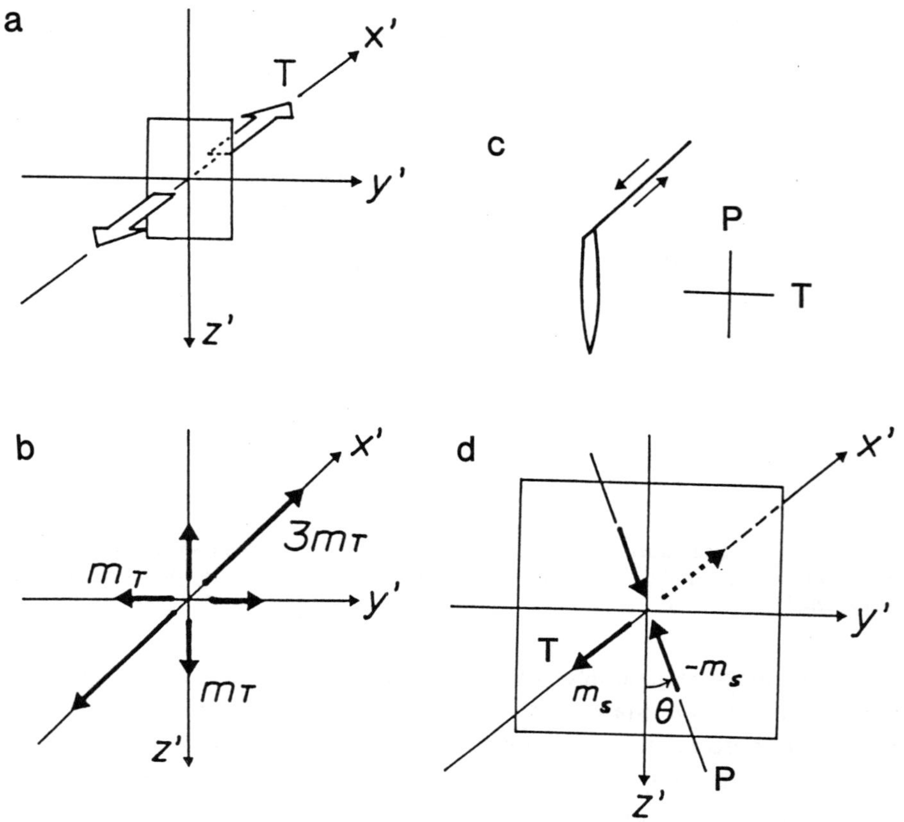

Figure 1 – 12. Moment tensors for a tensile crack and for a shear crack. (a) A tensile crack dislocates in x' direction. T is the tensile stress. (b) The body force equivalent to a tensile crack, where $3m_T$ is the tensile seismic moment. (c) A shear crack associated with the opening tensile crack. T and P indicate tension and compression axes of the shear crack, respectively. The shear crack is considered to have the same tension axis as the tensile crack. For sudden closing of an opened crack, the sign of forces should be reversed. (d) The body force equivalent to the shear crack. θ is the angle between P and z' axes on the $y'z'$ plane. m_S is the seismic moment of one of the double-couple forces.

1.5.1. *Moment Tensor of Tensile Shear Crack*

The moment tensor $\boldsymbol{M}$ in (1-26) is replaced by $\boldsymbol{M}_{TS}$ for a tensile-shear crack, represented as the sum of the moment tensor elements for both a tensile crack and a shear crack. For a tensile crack on the $y'z'$ plane with dislocation in the x' direction (Fig. 1-12), the moment tensor $\boldsymbol{M}_T$ in (x', y', z') coordinates is

$$\boldsymbol{M}_T = \begin{pmatrix} 3m_T & 0 & 0 \\ 0 & m_T & 0 \\ 0 & 0 & m_T \end{pmatrix} \qquad (1-28)$$

where $3m_T$ is the strength of tensile seismic moment. This tensile crack, shown in Fig. 1-12(a), is equivalent to the superposition of three vector dipoles, Fig. 1-12(b), if the crack is regarded as a point source.

Figure 1-12 is the illustration for an opening crack, and thus first motions of P-waves are compressional in all directions. Changing the sign of the vector dipoles in Fig. 1-12(b) gives a closing tensile crack which predicts dilatational P-wave initial motions.

The moment tensor for a shear crack is shown in Fig. 1-12(d). In this case, it is reasonable to assume that the tension axis T for the shear crack coincides with the maximum tension axis (x' axis) for the tensile crack. Therefore the compression axis P for the shear crack is on the $y'z'$ plane. Let θ be an angle between the z' axis and the compression axis. All these characterize the general force system of each volcanic earthquake. The moment tensor for the shear crack with this geometry is

$$\boldsymbol{M}_S = \begin{pmatrix} m_S & 0 & 0 \\ 0 & -m_S \sin^2\theta & -m_S \sin\theta\cos\theta \\ 0 & -m_S \sin\theta\cos\theta & -m_S\cos^2\theta \end{pmatrix}, \qquad (1-29)$$

where m_S is the seismic moment of the shear crack.

Consequently, the moment tensor for a tensile shear crack opening in x'-direction is

$$\boldsymbol{M}_{TS} = \boldsymbol{M}_T + \boldsymbol{M}_S. \qquad (1-30)$$

The same expressions can be also used for a closing tensile shear crack for which both m_T and m_S should have negative values.

1.5.2. *Focal Mechanism of Volcanic Earthquakes*

Seismic body waves radiated from a tensile shear crack can therefore be expressed in terms of the following five parameters:

(1) ϕ_N: azimuth angle of the normal of the tensile crack

(2) δ: plunge angle of the normal of the tensile crack

(3) $3m_T$: seismic moment for the tensile crack

(4) θ: angle between the compression axis and the dip direction of the tensile crack

(5) m_S: seismic moment for the shear crack

Generally, an inversion technique (which minimizes the squared residuals) is used to determine these unknown parameters for each volcanic earthquake by finding the best match between model calculations and observations of the amplitudes of P- and S-waves.

Figure 1-13 shows the azimuthal variation of observed and calculated amplitudes for four volcanic earthquakes. Synthetic amplitudes agree well

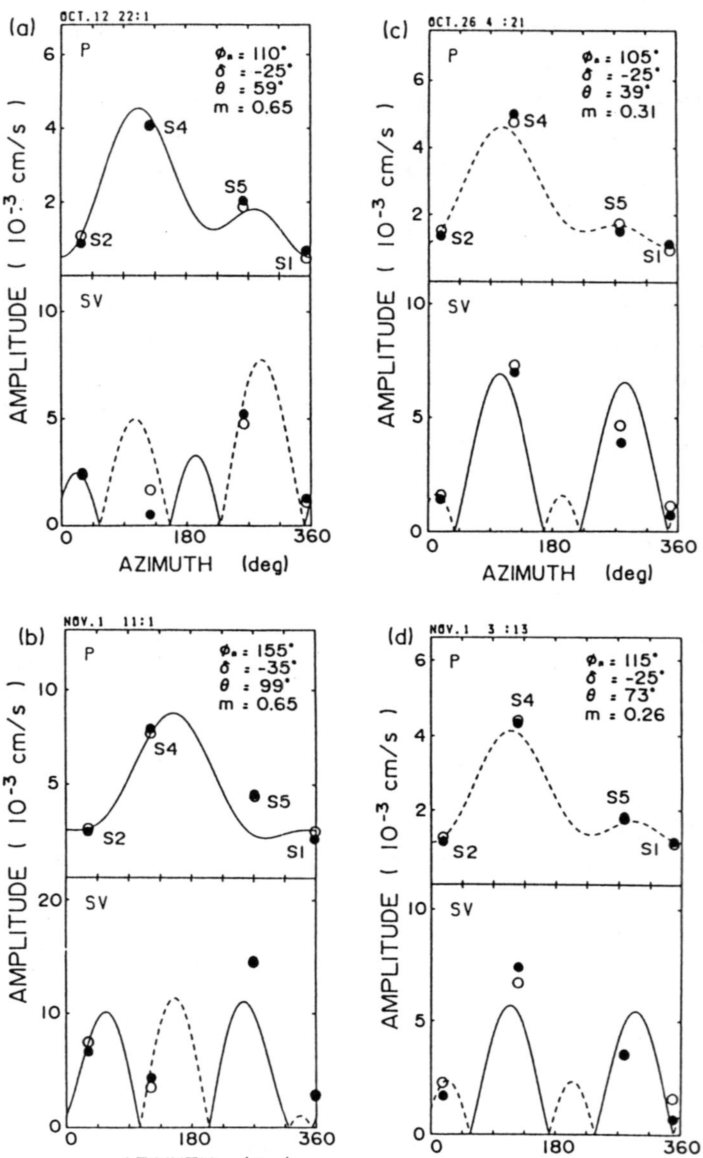

Figure 1 – 13. Amplitudes of P- and SV-waves as a function of azimuth (Shimizu et al., 1987). Solid circles indicate observed amplitudes. Open circles indicate values calculated from the tensile shear crack model. Solid and dashed curves are positive and negative amplitude patterns calculated by using the take-off angle of S2 station. Parameter m is the ratio m_S/m_T, which indicates the relative strength of the shear crack to the tensile crack. (a) and (b) are for compressional volcanic earthquakes on October 12 and November 1, 1983. (c) and (d) are for dilatational volcanic earthquakes on October 26 and November 1 following the 1983 Miyakejima eruption.

with observed amplitudes both for the compressional earthquakes (Figs. 1-13(a) and (b)) and for the dilatational earthquakes (Figs. 1-13(c) and (d)). This result is true not only for the four cases of earthquakes presented in Fig.1-13 but also for the other earthquakes analyzed. Therefore the tensile shear crack is an appropriate source model for volcanic earthquakes at Miyakejima volcano.

The ratio of strengths of the moments, $\frac{m_S}{m_T}$, is usually between from 0.3 and 1.0. Thus tensile cracking is dominant in the generation of volcanic earthquakes. This is why the same polarities of P-wave initial motions are observed at all stations, even though there is a shear component in the sources. It has been also shown that the strike of the tensile cracks is in good agreement with the direction of the surface fissures of the 1983 eruption. This also suggests the generation mechanism of the volcanic earthquakes due to the magma intrusion. In contrast to this, there was no preferred orientation of shear cracks.

1.6. SEISMIC WAVES FROM FINITE FAULTING

The displacement due to a double-couple point force is given in (1-12) for a seismic moment of $\mu D_0 \delta\Sigma$, where D_0 is the average amount of dislocation and $\delta\Sigma$ is the size of fault segment at $\vec{\xi}$. The displacement due to the general moment tensor is given in (1-13) and (1-26). By integrating the double-couple point force over a particular fault plane, the displacement due to finite faulting can be obtained.

Far-field displacement only is considered here. Compared with a fault segment at the origin, a fault segment at $\vec{\xi}$ from the origin is closer to an observation point by $\vec{\xi} \cdot \vec{\gamma}$, where $\vec{\gamma}$ is the unit vector pointing in the observation direction. Therefore, the moment rate function in (1-12) can be rewritten

$$\dot{M}_o(\vec{\gamma};t) = \mu \iint_S \dot{D}(\vec{\xi}, t + \frac{\vec{\xi} \cdot \vec{\gamma}}{c}) d\Sigma, \qquad (1-31)$$

where $\dot{D}(\vec{\xi}, t)$ is the dislocation velocity on the fault segment at $\vec{\xi}$, S is the size of the fault, and c stands for P or S-wave velocity of α or β. The moment rate tensor in (1-26) can be similarly rewritten.

The Fourier transform of (1-31) with respect to t is

$$\Omega(\vec{\gamma};\omega) = \mu \int_{-\infty}^{\infty} \exp(-i\omega t) dt \iint_S \dot{D}(\vec{\xi}, t + \frac{\vec{\xi} \cdot \vec{\gamma}}{c}) d\Sigma, \qquad (1-32)$$

where ω is angular frequency. As ω tends to zero, the Fourier transform $\Omega(\vec{\gamma};\omega)$ approaches a constant value:

$$\begin{aligned}\Omega(\vec{\gamma};\omega\to 0) &= \mu\int_{-\infty}^{\infty}dt\iint_S \dot{D}(\vec{\xi},t+\frac{\vec{\xi}\cdot\vec{\gamma}}{c})d\Sigma \\ &= \mu\iint_S D(\vec{\xi},t\to\infty)d\Sigma \\ &= \mu D_0 S\ (=M_o). \qquad (1-33)\end{aligned}$$

The right hand side of the above equation is the seismic moment of the finite faulting.

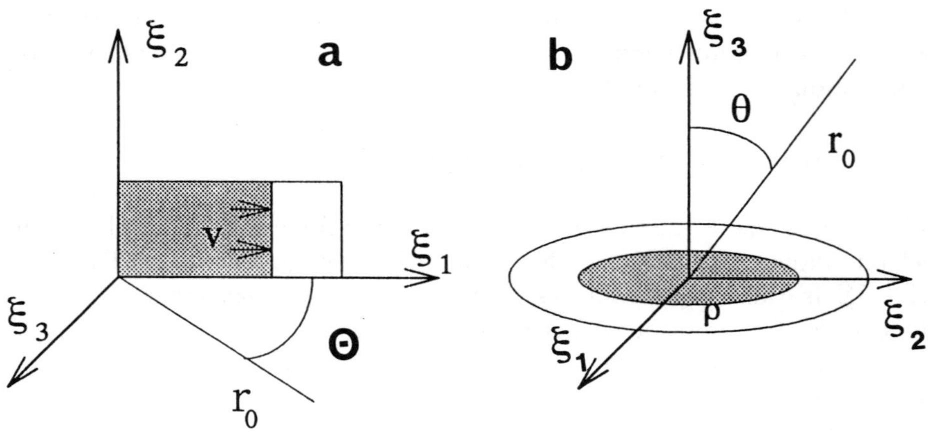

Figure 1 – 14. Geometry for a rectangular fault and a circular fault. (a) Unidirectional faulting on the $\xi_1\xi_2$ plane with fault length of extent L and width W. Observation is in the direction of Φ measured from the ξ_1 axis. v is the rupture velocity. (b) Circular rupture propagation on the $\xi_1\xi_2$ plane. The rupture starts from the origin and spreads with a constant rupture velocity v. This circular faulting is symmetric about the ξ_3 axis and the observation point is measured from the ξ_3 axis with angle θ.

1.6.1. *Source Spectrum from Rectangular Fault*

Considered a rectangular fault with length L and width W. Let the coordinate system (ξ_1,ξ_2) be parallel to the length and width of the fault plane as shown in Fig. 1-14. For a step function of dislocation with time, the moment rate function in (1-31) becomes

$$\dot{M}_o(\vec{\gamma};t-\frac{r_0}{c}) = \mu\int_0^W\int_0^L D(\xi_1,\xi_2)\delta(t-\frac{r_0}{c}-\frac{\xi_1}{v}+\frac{\xi_1\gamma_1+\xi_2\gamma_2}{c})d\xi_1 d\xi_2, \qquad (1-34)$$

where δ is the Dirac delta function resulting from the time derivative of the step function of dislocation and v is rupture velocity. The term $\frac{\xi_1}{v}$ represents the arrival time of the rupture front at ξ_1, when the dislocation starts on that fault segment. Assuming that W and $\xi_2\gamma_2$ are small, we can rewrite (1-34) as

$$\dot{M}_o(\vec{\gamma}; t - \frac{r_0}{c}) = \mu W \int_0^L D(\xi_1)\delta\Big(t - \frac{r_0}{c} - \xi_1\Big[\frac{1}{v} - \frac{\cos\Theta}{c}\Big]\Big) d\xi_1, \quad (1-35)$$

where Θ is the angle between the direction to the receiver and the direction of rupture propagation (ξ_1 direction).

For a constant spatial dislocation weighting function

$$D(\xi_1) = D_0, \quad (1-36)$$

the Fourier transform Ω of the moment rate function (1-35) is

$$\begin{aligned} \Omega(\vec{\gamma}; \omega) &= \mu D_0 W \int_0^L \exp\Big(-i\omega\xi_1\Big[\frac{1}{v} - \frac{\cos\Theta}{c}\Big]\Big) d\xi_1 \\ &= \mu D_0 W L \frac{\sin X}{X} \exp\Big(i\Big[\frac{\omega r_0}{c} - \frac{\pi}{2} + X\Big]\Big), \end{aligned} \quad (1-37)$$

where

$$X = \frac{\omega L}{2}\Big(\frac{1}{v} - \frac{\cos\Theta}{c}\Big). \quad (1-38)$$

The amplitude spectrum, which is the absolute value of (1-37), is designated the source spectrum.

The dislocation time function may be characterized by a rise time T_0 rather than being a step function. For example, a ramp time function is

$$D(t) = \begin{cases} 0, & t \le 0; \\ \frac{D_0}{T_0}\, t, & 0 < t \le T_0; \\ D_0, & T_0 < t. \end{cases} \quad (1-39)$$

In this case the Fourier transform Ω of the moment rate function in (1-35) can be similarly calculated as for (1-37). The amplitude spectrum of Ω is given

$$\begin{aligned} |\Omega(\vec{\gamma}; \omega)| &= \mu D_0 W L \Big|\frac{\sin X}{X}\Big| \Big|\frac{1 - \exp(-i\omega T_0)}{\omega T_0}\Big| \\ &= \mu D_0 W L \Big|\frac{\sin X}{X}\Big| \Big|\frac{\sin T_\tau}{T_\tau}\Big|, \end{aligned} \quad (1-40)$$

where

$$T_\tau = \frac{\omega T_0}{2}. \qquad (1-41)$$

Since $\frac{\sin X}{X} \to 1$ and $\frac{\sin T_\tau}{T_\tau} \to 1$ as $\omega \to 0$, the amplitude spectra of (1-37) and (1-40) both approach the seismic moment. We have seen this in (1-33).

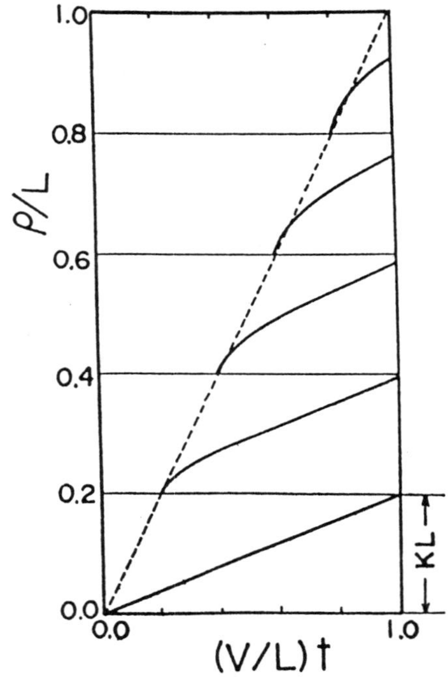

Figure 1 – 15. Dislocation time function on the circular fault (Sato and Hirasawa, 1973). The dashed line indicates the arrival of rupture front at each point ρ from the center normalized by the final radius of the circular fault L . Dislocation increases as shown by solid curve. $\rho/L = 0$ is the center, where the dislocation increases in proportion to vt as much as KL. $\rho/L = 1$ is the edge of the fault (the rupture terminates), where there is no dislocation.

For high frequencies $\omega \gg 1$, the term $\frac{\sin X}{X}$ behaves as ω^{-1}. Therefore, the spectral amplitude decreases with frequency as ω^{-1} in the case of the step time function of dislocation in (1-37) and as ω^{-2} in the case of the ramp time function in (1-40). Thus both the finite length of the faulting and the finite rise time of the dislocation cause the spectral amplitude to decrease at high frequencies.

1.6.2. *Source Spectrum from Circular Fault*

Rectangular faults propagating unidirectionally are too simplified to describe the real faulting of earthquakes. They may be appropriate only for very large earthquakes, where fault length L is much larger than fault width W. For a relatively small earthquake, it may be more appropriate to consider a rupture which initiates at a point and spreads radially at uniform speed, forming a circular fault surface.

Suppose that rupture on the $\xi_1\xi_2$ plane initiates at $t = 0$ and spreads from the origin radially with a uniform velocity v. The rupture front at a time t is a circle with a radius of vt as shown in Fig. 1-14b. When the relative dislocation at each instant is specified by the static solution of a circular crack under a uniform shear stress, the spatial dislocation function at ρ from the center is

$$D(\rho, t) = K\{(vt)^2 - \rho^2\}^{1/2} H(t - \frac{\rho}{v}), \qquad (1-42)$$

where $K = \dfrac{24}{7\pi}\dfrac{\Delta\sigma_0}{\mu}$ with applied shear stress $\Delta\sigma_0$ and rigidity μ, and $H(t)$ is Heaviside step function. Assuming that the faulting stops abruptly at the time when the rupture front reaches to L, the dislocation time function at ρ from the center of the circular fault is specified by

$$D(\rho, t) = \begin{cases} K\{(vt)^2 - \rho^2\}^{1/2} H(t - \frac{\rho}{v})\{1 - H(\rho - L)\}, & 0 < vt \le L; \\ K\{L^2 - \rho^2\}^{1/2}\{1 - H(\rho - L)\}, & vt > L. \end{cases} \qquad (1-43)$$

This model is constructed from the static solution of rupture formation for a circular crack and not from the solution for dynamical stress relaxation on the fault plane. However, the dislocation in (1-43) represents a first order approximation to the dynamical faulting process. The dislocation time function in (1-43) is illustrated in Fig. 1-15. The duration of the slip time function is dependent on the distance ρ from the center. The center of the fault slips for a longer time than the edges and a larger dislocation takes place at the center.

Putting the dislocation time function (1-43) into the moment rate function of (1-31), we obtain the following result:

$$\dot{M}_o(\theta; t) = \begin{cases} 2\mu K v L^2\left\{\dfrac{\pi x^2}{(1-k^2)^2}\right\}, & \text{for } 0 < x \le 1-k; \\ 2\mu K v L^2(\dfrac{\pi}{4})\left\{\dfrac{1}{k} - \dfrac{x^2}{k(1+k)^2}\right\}, & \text{for } 1-k < x \le 1+k, \end{cases} \qquad (1-44)$$

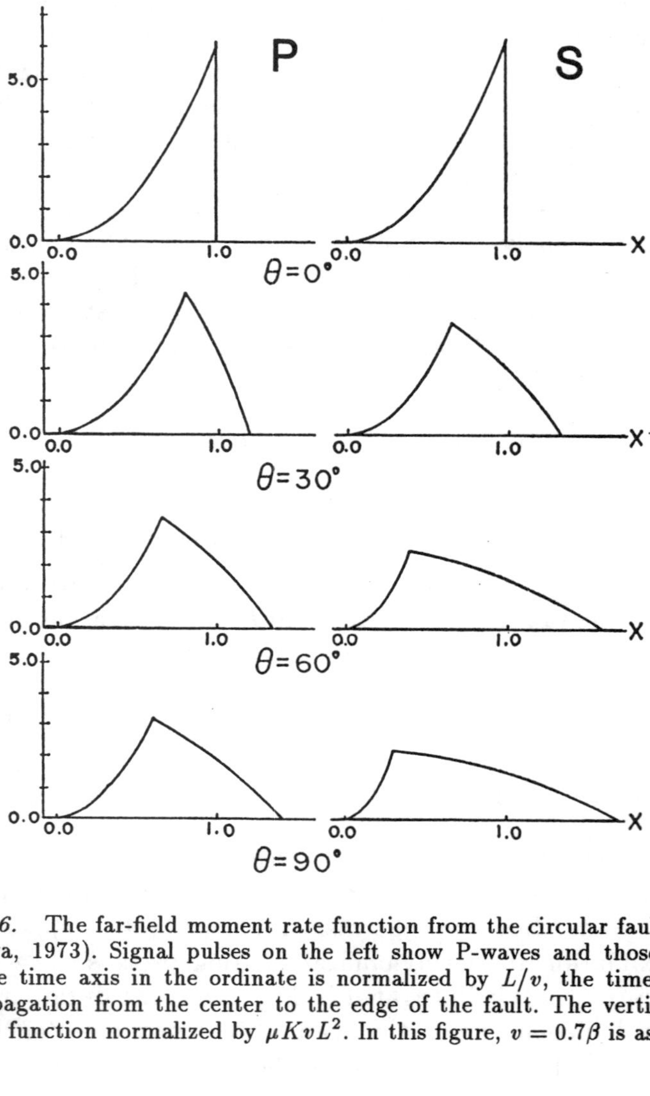

Figure 1 – 16. The far-field moment rate function from the circular fault model (Sato and Hirasawa, 1973). Signal pulses on the left show P-waves and those on the right S-waves. The time axis in the ordinate is normalized by L/v, the time necessary for rupture propagation from the center to the edge of the fault. The vertical axis is the moment rate function normalized by $\mu K v L^2$. In this figure, $v = 0.7\beta$ is assumed.

where $k = \dfrac{v}{c}\cos\theta$ (c is P- or S-wave velocity) and the time is changed to a dimensionless quantity $x = \dfrac{v}{L}(t - \dfrac{r_0}{c})$. Since the circular fault plane is symmetric, the unit vector pointing to the observation can be expressed by polar angle θ only. Figure 1-16 shows the moment rate function at different angles of θ. Signals in Fig. 1-16 are characterized by an initial quadratic increase with duration time corresponding to the growth of the rupture, followed by an abrupt decrease at the time of arrival from the rupture termination. The far field displacements of P- and S-waves can be evaluated through (1-12) using (1-44).

The source spectrum of this circular crack model can be calculated analytically as

$$\begin{aligned}
\Omega(\theta;\omega) &= \frac{3M_o}{\omega_L^2 k(1-k^2)}\Big[k\cos(\omega_L k)\cos\omega_L + \sin(\omega_L k)\sin\omega_L \\
&+\frac{1}{\omega_L(1-k^2)}\{(1+k^2)\sin(\omega_L k)\cos\omega_L - 2k\cos(\omega_L k)\sin\omega_L\}\Big] \\
&+\frac{i3M_o}{\omega_L^2 k(1-k^2)}\Big[\sin(\omega_L k)\cos\omega_L - k\cos(\omega_L k)\sin\omega_L \\
&+\frac{1}{\omega_L(1-k^2)}\{2k-(1+k^2)\sin(\omega_L k)\sin\omega_L \\
&-2k\cos(\omega_L k)\cos\omega_L\}\Big], \qquad (1-45)
\end{aligned}$$

where M_o is the seismic moment

$$M_o = \frac{16}{7}\Delta\sigma_0 L^3, \qquad (1-46)$$

and the angular frequency ω is normalized as $\omega_L = \left(\frac{L}{v}\right)\omega$.

Although the source spectrum (1-45) is complicated, it has constant amplitude at low frequencies corresponding to the seismic moment in (1-46). At high frequencies the amplitude asymptote decreases proportionally to ω^{-2}. This is shown in Fig. 1-17. This spectral behavior is similar to that of (1-40).

1.6.3. *Corner Frequency and Source Size*

A spectral corner frequency is defined by the frequency at which the high and low frequency trends of the source spectrum intersect. First consider rectangular faults with fault length L and width W. For the very narrow fault formulated in (1-37), the corner frequency is

$$\omega_c = \left\{\frac{L}{2v}\left(1-\frac{v}{c}\cos\Theta\right)\right\}^{-1}, \qquad (1-47)$$

because of the frequency dependence of $\frac{\sin X}{X}$. For a rise time T_0 of $\frac{W}{2v}$, a corner frequency may be obtained as a geometric mean of the two corner frequencies which are described in (1-40) and are associated with the finitely propagating rupture and the rise time. The corner frequency averaged over all directions for P-waves is

$$2\pi < f_P >_r = \frac{\sqrt{2.9}\alpha}{\sqrt{LW}}, \qquad (1-48)$$

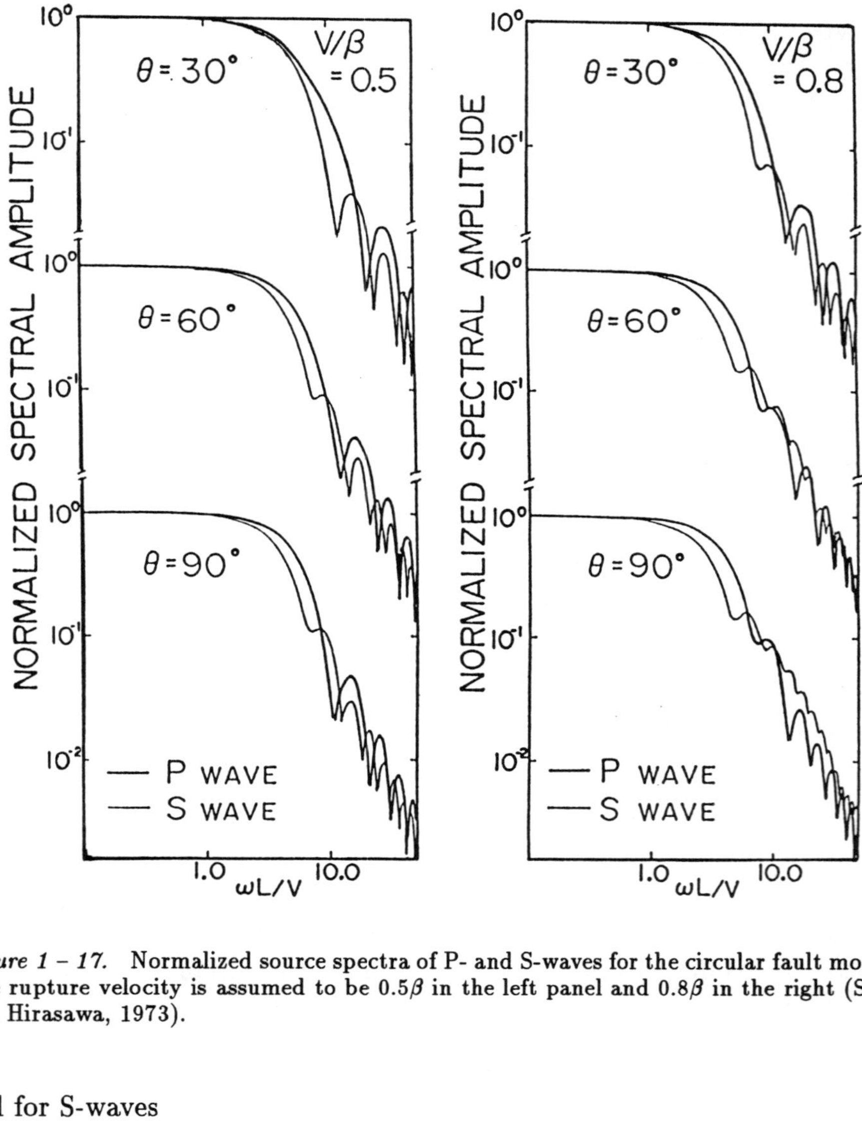

Figure 1 – 17. Normalized source spectra of P- and S-waves for the circular fault model. The rupture velocity is assumed to be 0.5β in the left panel and 0.8β in the right (Sato and Hirasawa, 1973).

and for S-waves

$$2\pi < f_S >_r = \frac{\sqrt{14.8}\beta}{\sqrt{LW}}, \qquad (1-49)$$

where the rupture velocity is assumed $v = 0.9\beta$. It should be noted that the corner frequency of S-waves described in (1-49) is larger than that of P-waves in (1-48), when Poisson's ratio is 0.25.

The corner frequency for circular cracks varies with θ. An averaged value of corner frequencies over all directions for P-waves can be calculated as

$$2\pi < f_P >_c = \frac{C_P \alpha}{L}, \qquad (1-50)$$

and for S-waves

$$2\pi < f_S >_c = \frac{C_S \beta}{L}, \qquad (1-51)$$

where C_P and C_S are 1.43 and 1.76, when $v = 0.8\beta$, and 1.53 and 1.85, respectively, when $v = 0.9\beta$. It is noteworthy that the P-wave corner frequency is larger than that of S-waves at every direction in this case of the circular fault except only for $\theta = 0$ where the spectra for P- and S-waves are identical.

The relations between fault length (or radius) L and corner frequency ω_c from (1-47) to (1-51) are very important, because the corner frequency has been estimated from observed body-waves after correcting for the effects of wave travel distances. Many studies since the end of 1960's have tried to obtain the source extent from the Fourier spectra of observed body-waves. However, such estimates of earthquake source sizes depend significantly on the assumption of the source models and on physical parameters such as rupture velocity. Furthermore, the difference in ω_c's for P- and S-waves causes the drastic change in energy partition between P- and S-waves. Therefore, the evaluating corner frequencies is very important not only to estimate the earthquake source size but also to understand the energy budget and the appropriate model for the earthquake source process.

1.7. QUANTIFICATION OF EARTHQUAKES

Seismic moment M_o is one of the important measures of earthquake strengths. Recent advances in seismometry and computer facilities enable determination of this source parameter of earthquakes in a routine manner. Figure 1-18 shows the result of moment tensor inversion for major earthquakes in the world (magnitude larger than 5) on June, 1994. Moment tensor and/or centroid hypocenter for each earthquake are determined from long-period body-waves and long-period wave forms at teleseismic distances. Centroid hypocenter is the position where the dominant seismic energy of a particular earthquake is radiated, whereas common hypocenter is the point where the initial breakage takes place. Table 1-1 lists the results for the first three events which occurred in June, 1994. The table includes origin time, hypocenter, focal mechanism solution and seismic moment from body-waves, and centroid hypocenter, focal mechanism solution and seismic moment from long-period wave forms. These results are reported routinely by the mail service of the U. S. Geological Survey and issued periodically.

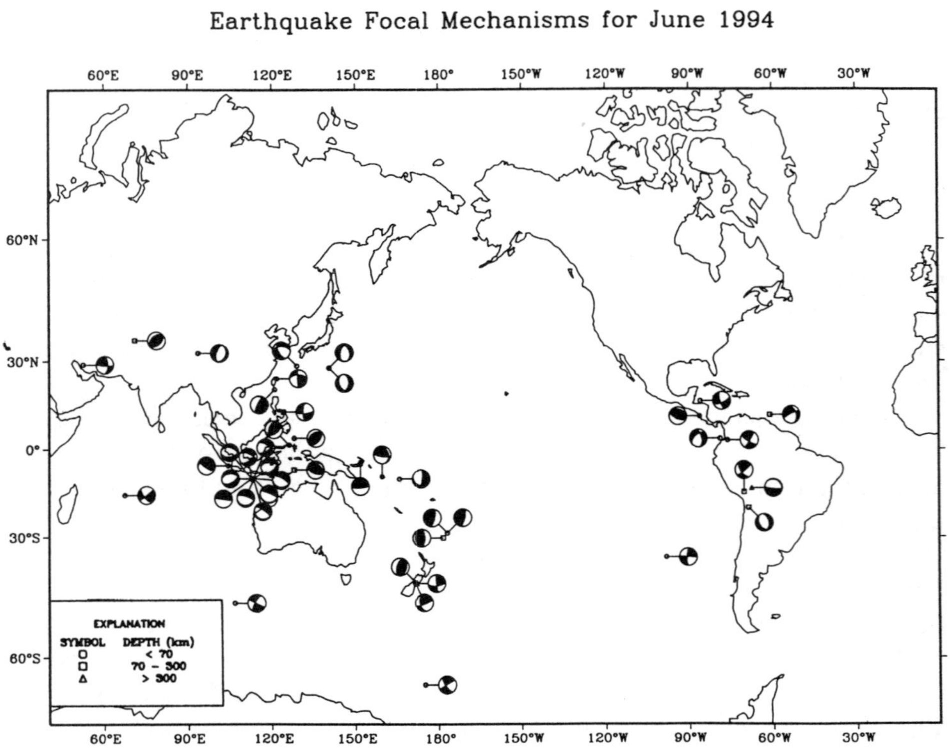

Figure 1 – 18. **Earthquake focal mechanisms of major earthquakes in June, 1994 (U. S. Geological Survey, 1994). Source parameters of the first three earthquakes are listed in Table 1-1.**

Earthquake magnitude has been used as a measure of the strength of earthquakes. Local magnitude or Richter scale, M_L, was introduced to measure the strength of shallow earthquakes in California. Allowing for the nominal magnification of the Wood-Anderson seismograph and applying an empirical formula for southern California to correct wave travel distances, M_L is calculated from

$$M_L = \log A + 2.56 \log \Delta - 1.67, \qquad (1-52)$$

where A is the maximum amplitude of ground motion in μm and Δ is epicentral distance in km. This magnitude scale was first applied to local earthquakes within 600km.

Seismograms of shallow earthquakes usually show dominant surface-waves at a period of about 20 sec at teleseismic distances. Surface-wave

A D D I T I O N A L   S O U R C E   P A R A M E T E R S

```
01 03 13 47.15  11.913N  60.998W  73km
   5.3mb ( 98 obs.)
   WINDWARD ISLANDS
   CENTROID, MOMENT TENSOR      (HRV)
   Data Used: GDSN
   L.P.B.: 31S, 41C
   Centroid Location:
   Origin Time        03:13:48.2 0.4
   Lat 11.75N 0.07 Lon  60.64W 0.05
   Dep  51.9 3.8 Half-duration  2.5
   Principal Axes:
     Scale 10**16 Nm
     T  Val= 17.14  Plg=22  Azm=300
     N       -2.48      44       53
     P      -14.67      38      192
   Best Double Couple:Mo=1.6*10**17
    NP1:Strike=343 Dip=45 Slip=-166
    NP2:        243     80      -46

02 04 39 35.54  20.951N 121.137E  23km
   5.1mb ( 64 obs.)  4.9Msz (  6 obs.)
   PHILIPPINE ISLANDS REGION
   CENTROID, MOMENT TENSOR      (HRV)
   Data Used: GDSN
   L.P.B.: 24S, 34C
   Centroid Location:
   Origin Time        04:39:39.6 0.3
   Lat 21.05N 0.06 Lon 121.23E 0.06
   Dep  15.0 FIX Half-duration  1.3
   Principal Axes:
     Scale 10**16 Nm
     T  Val= 14.17  Plg=59  Azm=135
     N       -0.56      13       22
     P      -13.61      28      285
   Best Double Couple:Mo=1.4*10**17
    NP1:Strike=344 Dip=21 Slip=  50
    NP2:        206     74      104
```

```
02 18 17 34.02  10.477S 112.835E  18km
   5.7mb ( 52 obs.)  7.2Msz ( 37 obs.)
   SOUTH OF JAWA, INDONESIA
   FAULT PLANE SOLUTION: P-Waves
    NP1:Strike= 98 Dip=85 Slip=  90
    NP2:        278      5        90
   Principal Axes:
     T                Plg=50  Azm=  8
     P                    40      188
   Comment: The focal mechanism is
     poorly controlled and
     corresponds to reverse
     faulting. The preferred fault
     plane is NP2.
   RADIATED ENERGY
    No. of sta:  24  Focal mech.  F
    Energy        1.2±0.2*10**14 Nm
   MOMENT TENSOR SOLUTION
   Dep   6             No. of sta: 31
   Principal Axes:
     Scale 10**20 Nm
     T  Val=  5.68  Plg=43  Azm=356
     N       -0.93      11       96
     P       -4.75      45      198
   Best Double Couple:Mo=5.2*10**20
    NP1:Strike= 11 Dip=11 Slip=-175
    NP2:        276     89      -79
   CENTROID, MOMENT TENSOR     (HRV)
   Data Used: GDSN
                       M.W.: 56S,151C
   Centroid Location:
   Origin Time        18:18:15.8 0.1
   Lat 11.03S 0.01 Lon 113.04E 0.01
   Dep  15.0 FIX Half-duration 11.5
   Principal Axes:
     Scale 10**20 Nm
     T  Val=  5.43  Plg=52  Azm=  9
     N       -0.18       0      279
     P       -5.26      38      189
   Best Double Couple:Mo=5.3*10**20
    NP1:Strike=278 Dip= 7 Slip=  89
    NP2:         99     83       90
```

TABLE 1 – 1. Moment tensor inversion of major earthquakes in June, 1994 by the U. S. Geological Survey.

magnitude M_S is defined for shallow-focus earthquakes as

$$M_S = \log A + 1.656 \log \Delta + 1.818, \qquad (1-53)$$

where Δ is epicentral distance in degree, and A is the combined maximum amplitude on horizontal components in micron. Surface-wave magnitude for major earthquakes in the world has been determined since the end of the 19th century.

By extending the empirical tables to cover significant ranges in focal depth and distance, body waves in teleseismic distances have also been analyzed to measure the strength of both shallow and deep earthquakes.

Body-wave magnitude m_B is calculated from

$$m_B = \log\left(\frac{A}{T}\right) + Q(\Delta, H), \tag{1-54}$$

where A is the maximum amplitude of P-waves on the vertical or horizontal component, and or S-waves on the horizontal component, T is an apparent period of body waves in second. $Q(\Delta, H)$ is a calibration function correcting for the wave travel distance as a function of the epicentral distance Δ and the focal depth H of the earthquake.

Although the concept of surface-wave and body-wave magnitudes is unchanged, the historical development of seismological instruments gave rise to some differences in the magnitude scales. Surface-wave magnitude M_s reported currently by the U.S. Geological Survey(USGS) and the International Seismological Centre(ISC) is defined as

$$M_s = \log\left(\frac{A}{T}\right) + 1.66 \log \Delta + 3.3, \tag{1-55}$$

where A is the combined maximum amplitude of surface-waves at the period of about 20 sec on horizontal components or the maximum amplitude on the vertical component, both in micron. It has been reported that this definition of surface-wave magnitude causes an overestimate of about 0.2 unit larger than M_S of the same earthquake.

The body-wave magnitude m_b determined routinely by the USGS and the ISC is substantially different from m_B. This is mainly because m_b is evaluated from the maximum amplitude of P-waves within the initial few cycles of waves observed by narrow-band short-period seismometers, whereas m_B has been obtained from data recorded by broad-band instruments. It will be shown in §3 that the initial few cycles of P-waves do not represent the whole strength of large and great earthquakes.

Seismic energy is another measure to evaluate the strength of earthquakes. An empirical relation between radiated seismic energy and surface-wave magnitude M_S has been obtained as

$$\log E_S = 1.5 M_S + 11.8, \tag{1-56}$$

where E_S (in ergs) is calculated from seismic waves by a crude approximation. This enables us to estimate the seismic energy from the surface-wave magnitude M_S of an earthquake. Recent advances in digital seismometry brings us more precise estimates of the seismic energy of earthquakes. An extensive analysis will be made in §4, 5 and 6 to understand the energy budget of natural earthquakes by considering the complex faulting process.

CHAPTER 2

EARTHQUAKE SOURCE SPECTRUM OF COMPLEX FAULTING PROCESS

2.1. SOURCE SPECTRUM OF HETEROGENEOUS FAULTING

An earthquake fault is defined by a dislocation distribution $D(\xi, t)$ at points ξ on the plane at times t. In this chapter we investigate heterogeneous faulting represented by a highly random function of $D(\xi, t)$. Suppose that the fault plane extends for a length L and a width W within an infinite homogeneous medium. For simplicity, the faulting rupture is assumed to propagate unilaterally along the ξ axis for a distance L. Figure 2-1 illustrates a heterogeneous fault where a smooth and coherent rupture extends over a rectangular fault plane. In addition, non-uniform and incoherent ruptures propagate over localized small areas on the fault.

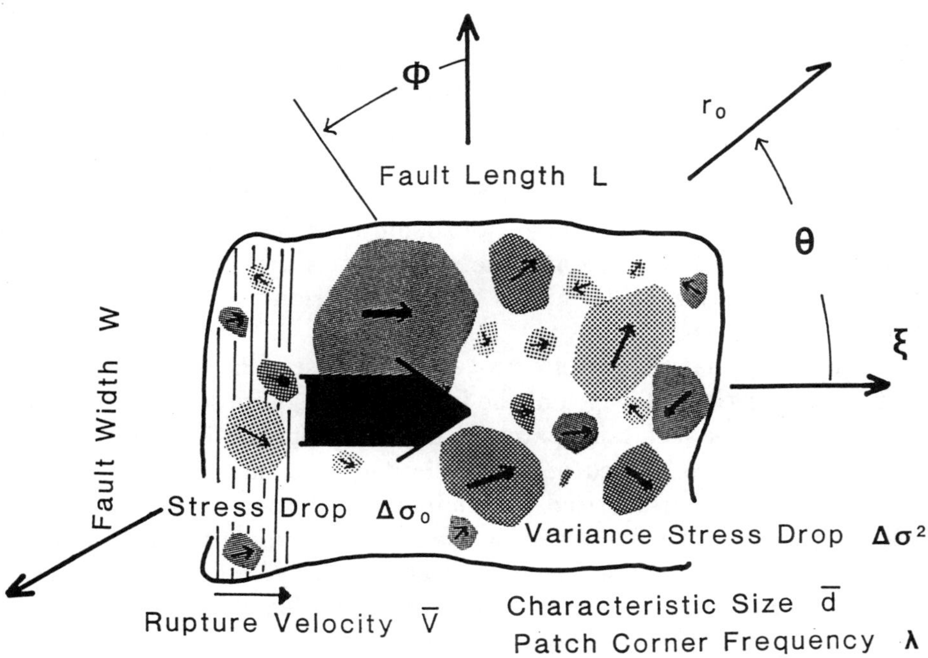

Figure 2 - 1. Schematic illustration of a heterogeneous earthquake fault where the dislocation across the fault is generated. Fault length and width, stress drop, and average dislocation are macroscopic source parameters of the complex faulting process. Random fault patches represent small-scale fault heterogeneities. Patch corner frequency and variance stress drop are stochastic source parameters of the complex faulting process (Koyama, 1994). Polar coordinates are defined as shown.

The displacement component of P-waves in the far-field from this kind of a shear fault is generally written as

$$u_r = \frac{(\beta/\alpha)^3}{4\pi\beta r_0} \sin 2\theta \sin\phi I_\alpha(t), \tag{2-1}$$

and those of S-waves as

$$\begin{aligned} u_\theta &= \frac{1}{4\pi\beta r_0} \cos 2\theta \sin\phi I_\beta(t), \\ u_\phi &= \frac{1}{4\pi\beta r_0} \cos\theta \cos\phi I_\beta(t), \end{aligned} \tag{2-2}$$

with

$$I_c(t) = W \int_0^L \dot{D}(\xi, t - \frac{r - \xi\cos\theta}{c}) d\xi, \tag{2-3}$$

where $\dot{D}(\xi, t)$ is a dislocation velocity time function averaged along the fault width on the fault plane. In the above equations, the distance of an observer is r_0 from the center of the fault, α and β indicate P- and S-wave velocities, where c stands for either of them. Polar coordinates (r, θ, ϕ) are introduced in Fig. 2-1: θ is polar angle of an observer measured from the fault strike, and ϕ is azimuth angle. Although we should consider more rigorously the dislocation velocity function and the rupture propagation effect along the fault width, equation (2-3) suffices as an expression for a very long fault without losing generality.

It is unlikely that the rupture process of small-scale heterogeneous areas is coherent over the full length of the fault. We designate such a small-scale heterogeneous area as a fault patch. It is more probable that fault patches are localized and correlated only over limited segments of the fault. The fault patches are characterized by random distribution of dislocations on localized areas of the heterogeneous fault. Therefore, the fluctuation of dislocations on this heterogeneous fault are correlated only over each fault patch. Figure 2-2 shows schematics of dislocation functions for a coherent rupture and for incoherent ruptures superposed on a coherent rupture.

Since an earthquake is essentially a transient phenomenon, the dislocation function $D(\xi, t)$ in Fig. 2-2 is zero when $t < 0$. Also it has constant value $D(\xi, \infty)$ for $t > T_0$, where T_0 is the time necessary to complete the dislocation on that part of the fault. Hereafter the heterogeneous fault is considered as a line source with length L and rise time T_0, being formulated in (1-40) and (2-3). The parameter T_0 is strictly related to a characteristic rupture time of the fault width as

$$T_0 = \frac{qW}{\bar{v}}, \tag{2-4}$$

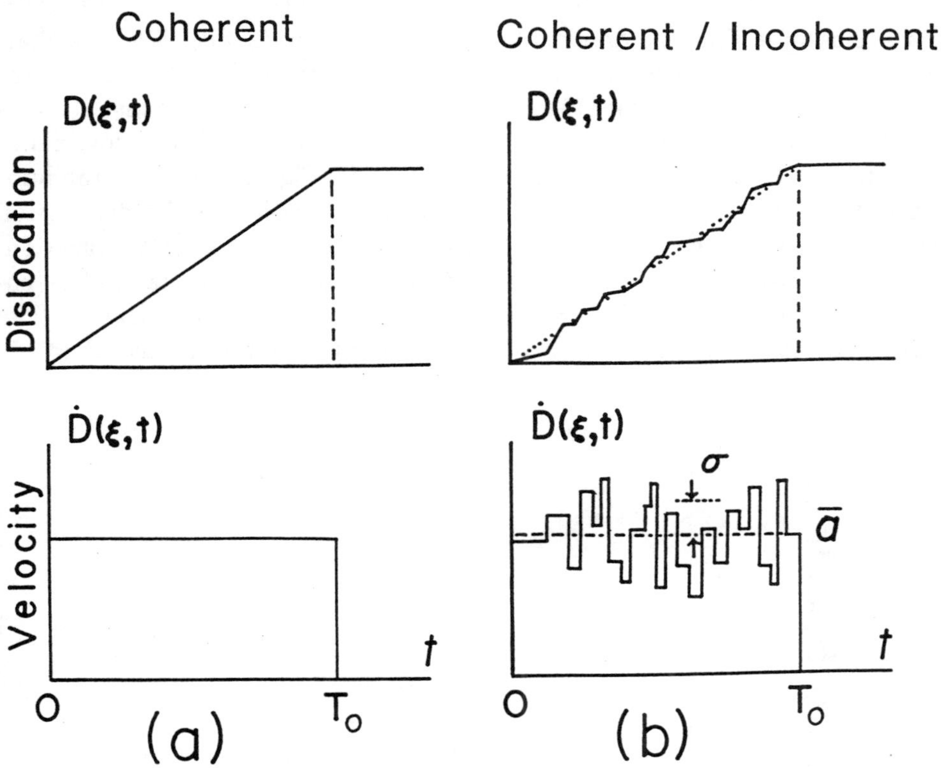

Figure 2 – 2. Dislocation time-function and dislocation velocity time-function at a point on a fault. Average dislocation velocity $\bar{a}$ and variance dislocation velocity σ^2 represent the deterministic and the stochastic parts of the complex faulting process (Koyama, 1985). The rise time of dislocation is T_0 and $\bar{v}$ is the average rupture velocity.

where $\bar{v}$ is an average rupture velocity and q is a constant. In the latter section, q is assumed to be $\frac{1}{2}$ irrespective of the earthquake source size.

For a coherent rupture, $\dot{D}(\xi, t)$ is represented by a dislocation velocity deterministically as in Fig. 2-2(a). The coherent rupture at ξ radiates a rectangular pulse with an amplitude of $\bar{a}$ and a rise time T_0, where $T_0\bar{a}$ is the average dislocation on the fault. On the other hand, if the faulting is highly irregular, $\dot{D}$ is described by the coherent part and incoherent part as in Fig. 2-2(b). The coherent rupture is the same as that in Fig. 2-2(a), characterized by the amplitude $\bar{a}$ and the time duration T_0. Also, the rupture propagation on a particular localized fault patch radiates a rectangular pulse. This is represented by the fluctuation of dislocation velocities superposed on $\bar{a}$. The dislocation velocity function in this case is not as simple as (2-3). The contribution from the coherent rupture can be evaluated deterministically by (2-3), but, that from the incoherent ruptures is formulated stochastically.

We assume that the amplitude of random pulses in Fig. 2-2(b) is characterized by a variance of dislocation velocities σ^2, and that pulse widths relate to the rupture propagation time on the fault patches. The statistical theory of communication has been used to obtain an autocovariance function of the dislocation velocity function in Fig. 2-2(b) as a random-pulse time series. Appendix A describes the derivation of the dislocation velocity function, considering a stochastic process with a finite duration. It is expressed as the squared sum of the smooth rectangular pulse (coherent rupture) and random rectangular pulses (incoherent ruptures). The autocovariance function of $\dot{D}(\xi,t)$ is expressed (Appendix A) as

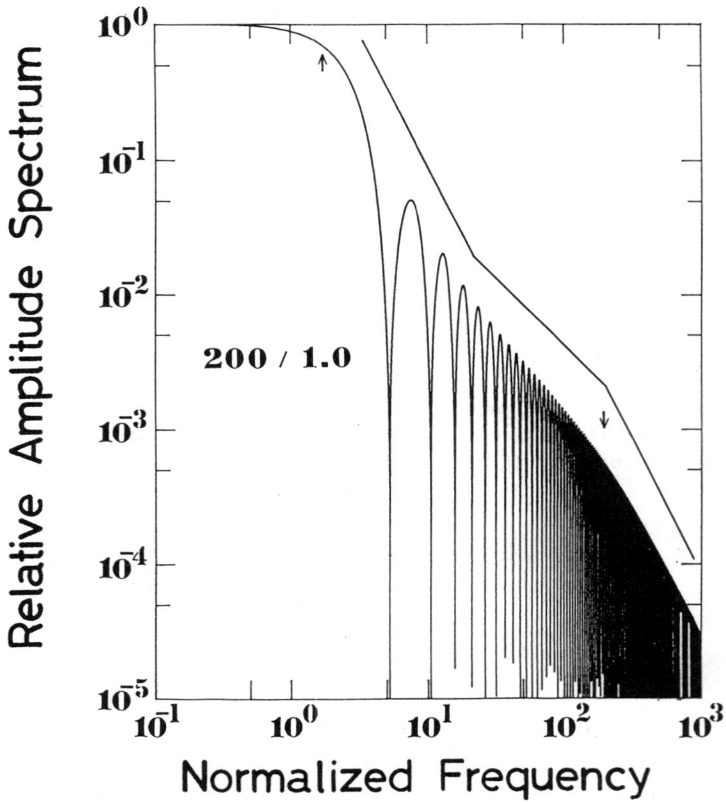

Figure 2 – 3. Source spectrum of the complex faulting process. Frequency in the ordinate is angular frequency normalized by W/β, where W is fault width and β is S-wave velocity. A rise time T_0 of $1.2(W/\beta)$, $L = W$, $\sigma^2/\bar{a}^2 = 1$, and patch corner frequency λ of $200/(W/\beta)$ are assumed. This spectrum shows a frequency dependence of $\omega^0 - \omega^{-2} - \omega^{-h} - \omega^{-2}$, where h is about 1.0. Two corner frequencies, where the spectral envelope changes its general trend, are found at about 1.7 and 200 (Koyama, 1985).

$$C(\tau) = \begin{cases} (T_0 - |\tau|)\ \{\bar{a}^2 + \sigma^2 \exp(-\lambda|\tau|)\}, & 0 \leq |\tau| < T_0; \\ 0, & T_0 \leq |\tau|, \end{cases} \tag{2-5}$$

where λ is a mean number of random-pulses in unit time. This is designated patch corner frequency and is related to

$$\lambda = \frac{\bar{v}}{\bar{d}}, \tag{2-6}$$

where $\bar{d}$ is the characteristic length of fault patches. This equation is understood to be the inverse of a mean rupture time of fault patches. Clearly, when $\sigma^2 \to 0$, $C(\tau)$ represents the autocovariance of a smooth rupture represented by a box-car function by $\bar{a}$ and T_0. The correlation of random pulses is restricted by the mean rupture time of fault patches, which appears in the second term within braces of (2-5).

The autocovariance in (2-5) is for a limited fault segment at ξ. As the heterogeneous faulting propagates, the faulting process is stationary in the time domain and also in the space domain; then the contribution of incoherent ruptures can be expressed similarly to (2-5). The autocovariance for the full-length of heterogeneous faulting is expressed by summing up (2-5) along the fault length L where the faulting rupture propagates at a constant velocity $\bar{v}$:

$$C(\xi;\tau) = \begin{cases} (L-|\xi|)(T_0 - |\tau - \frac{\xi}{\bar{v}}|)\{\bar{a}^2 + \sigma^2 \exp(-\lambda|\tau - \frac{\xi}{\bar{v}}|)\}, \\ \qquad 0 \leq |\xi| \leq L,\ 0 \leq |\tau - \frac{\xi}{\bar{v}}| \leq T_0; \\ 0, \qquad \text{otherwise.} \end{cases} \tag{2-7}$$

The source spectrum $|A_c(\omega)|$ of the complex faulting process can be expressed by the Fourier transform of the autocovariance function (Appendix A) as

$$\begin{aligned} |A_c(\omega)|^2 &= W^2 \left| \int_{-\infty}^{\infty} d\tau \exp(-i\omega\tau) \int_{-L}^{L} d\xi \exp(ik_c\xi)\ C(\xi;\tau) \right| \\ &= (LWT_0\bar{a})^2 \frac{\sin^2([k_c - \omega/\bar{v}]L/2)}{\{(k_c - \omega/\bar{v})L/2\}^2} \\ &\times \left[\frac{\sin^2(\omega T_0/2)}{(\omega T_0/2)^2} + \frac{2\sigma^2}{T_0^2 \bar{a}^2} \left\{ \frac{\omega^2 - \lambda^2}{(\omega^2+\lambda^2)^2} + \frac{\lambda T_0}{\omega^2+\lambda^2} \right\} \right], \end{aligned} \tag{2-8}$$

where k_c is the wave number of $\frac{\omega}{c}\cos\theta$, and ω is angular frequency, and the fault width is assumed much larger than the characteristic patch size,

$W >> \bar{d}$. Hereafter, all frequencies are angular frequency except for some special cases expressed by *frequency f*.

The displacement spectrum Ω of P- and S-waves radiated from this complex faulting process can be expressed by applying (2-8) to (2-1) and (2-2)

$$\Omega(P,S) = \frac{M_o}{4\pi\rho r_0(\alpha^3,\beta^3)}|R_{\theta\phi}(P,S)|\;|B_c(\omega)|, \tag{2-9}$$

$$|B_c(\omega)| = \frac{|\sin([k_c-\omega/\bar{v}]L/2)|}{|(k_c-\omega/\bar{v})L/2|}\left[\frac{\sin^2(\omega T_0/2)}{(\omega T_0/2)^2} + \frac{2\sigma^2}{T_0^2\bar{a}^2}\left\{\frac{\omega^2-\lambda^2}{(\omega^2+\lambda^2)^2}+\frac{\lambda T_0}{\omega^2+\lambda^2}\right\}\right]^{1/2}, \tag{2-10}$$

where M_o is seismic moment, and $R_{\theta\phi}(P,S)$ is the radiation pattern coefficient either for P- or S-waves from a double-couple point force.

Figure 2-3 shows an example of the normalized source spectrum for the parameters of $\frac{\sigma^2}{\bar{a}^2}=1$, $\lambda\left(\frac{W}{\beta}\right)=200$, $L=W$ and $\bar{v}\simeq 0.83\beta$. An observation direction $\theta=\frac{\pi}{2}$ is used to draw the example source spectrum in Fig. 2-3. The source spectrum in (2-8) shows a frequency dependence changing as $\omega^0-\omega^{-2}-\omega^{-h}-\omega^{-2}$ with increasing frequency ω. The power coefficient h is measured to be about 1. There appear two characteristic corner frequencies, where the spectral envelope changes its general trend; the first corner frequency results from the rupture propagation time of the entire faulting and the second corner frequency from the average rupture time of random fault patches. The former corner frequency has been investigated in (1-47). The latter corner frequency is strictly related to λ and is essential for characterizing the size effect due to the stochastic part of the complex faulting process.

The excitation of seismic body-waves can be represented by a seismic moment tensor as described in §1.4. The moment tensor appropriate for a general shear-fault is the combination of two point double-couples orthogonal each other, because the net moment of coupled forces must be zero for shear faultings. The moment tensor representation of the heterogeneous faulting may be obtained by adding displacement components from the other (minor) double-couple to (2-9). In order to make the discussion simple and suitable for actual applications, we do not proceed with the moment tensor representation of heterogeneous faulting.

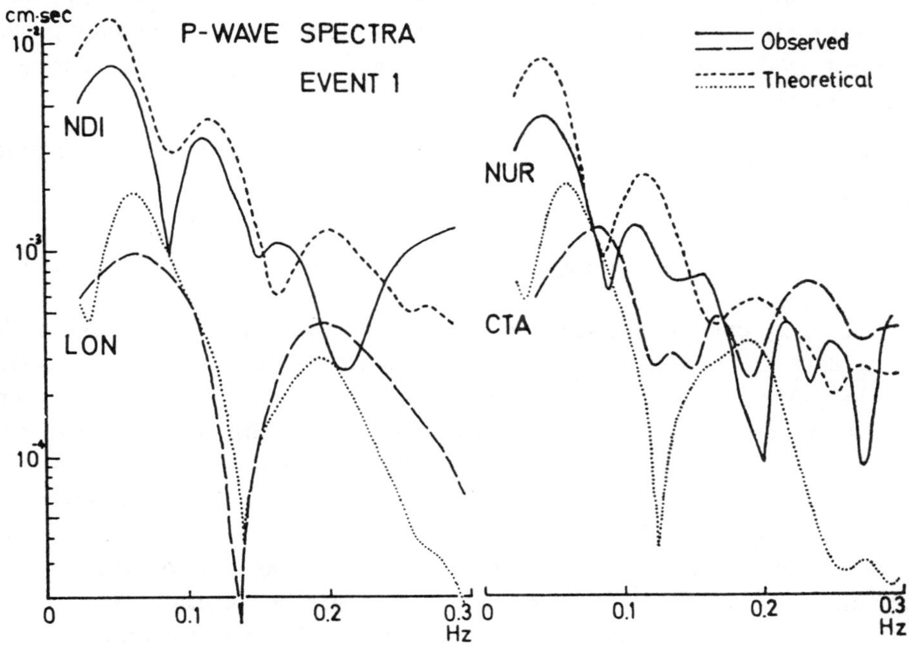

Figure 2 – 4. **Long-period P-wave spectra at observation stations of NDI, LON, NUR, and CTA for the 1965 Off Sanriku earthquake (M_s=5.6) on the northeastern Japan coast. Observed spectral troughs are fitted with those by the seismic Doppler effect (Izutani and Hirasawa, 1978).**

2.2. SOURCE PARAMETERS OF COMPLEX FAULTING PROCESS

2.2.1. *Seismic Doppler Effect and Corner Frequency*

The source spectrum in (2-8) is expressed in terms of the deterministic part and the stochastic part of the complex faulting process. The spectrum is characterized by a frequency dependence of $\left|\dfrac{\sin(\omega/\omega_c)}{\omega/\omega_c}\right|$ for the first term in (2-8). This has asymptotes $\simeq 1$ for $|\omega| < \omega_c$ and $\dfrac{1}{|\omega/\omega_c|}$ for $|\omega| > \omega_c$. The seismic moment of an earthquake in (1-33) can be obtained from observations of the low frequency limit after correcting for the constant coefficient and for the radiation pattern coefficient as described in (2-9). The source spectrum in (2-8) has a characteristic corner frequency of

$$\omega_c = \frac{2\bar{v}}{L(1 - \dfrac{\bar{v}}{c}\cos\theta)}, \qquad (2-11)$$

where the spectral envelope changes its general trend. This changes as a function of θ, the observation direction measured from the rupture propagation direction. This azimuthal dependence of the corner frequency is the seismic Doppler effect, resulting from the finitely propagating rupture in a particular direction.

The term $\left|\frac{\sin(\omega/\omega_c)}{(\omega/\omega_c)}\right|$ in the right hand side of (2-8) causes periodic zeros on the spectrum due to the destructive interference:

$$\omega = n\pi\omega_c \quad (n = \pm 1, \pm 2, ...). \tag{2-12}$$

Therefore, ω_c can be evaluated by measuring these periodic zeros on P- and S-wave spectra from observations. Figure 2-4 shows source spectra from long-period P-waves of an earthquake in Japan to demonstrate such spectral behavior. By matching a theoretical spectrum to the observed, rupture velocity $\bar{v}$ and fault length L can be evaluated. Observation of ω_c's at many stations at different azimuth angles enables us to estimate the rupture propagation direction of the faulting.

The seismic Doppler effect can be identified on long-period surface wave more directly than for body waves. This is because each station can observe surface waves which travel along both the minor great circle connecting the station and the epicenter of an earthquake and the major great circle. Corresponding station directions for these two waves are θ and $\theta + \pi$, respectively from the rupture propagation direction on the fault.

A seismic directivity function is defined by the spectral ratio of these surface waves as

$$F(\omega) = \left|\frac{\sin\left(\frac{\omega L}{2\bar{v}}[1 - \frac{\bar{v}}{c}\cos\theta]\right)}{1 - \frac{\bar{v}}{c}\cos\theta}\right| \Big/ \left|\frac{\sin\left(\frac{\omega L}{2\bar{v}}[1 + \frac{\bar{v}}{c}\cos\theta]\right)}{1 + \frac{\bar{v}}{c}\cos\theta}\right|. \tag{2-13}$$

This function can be evaluated from a single-station observation of surface waves. Figure 2-5 presents the seismic directivity function thus obtained at Pasadena, California for the case of the 1952 Kamchatkan earthquake. Peak frequencies and trough frequencies of the directivity function yield an estimate of 700km for L and 3km/sec for $\bar{v}$, and a rupture propagation direction of N140°E.

2.2.2. *Kinematical and Dynamical Source Parameters*

A source spectrum for coherent rupture is obtained, when the variance dislocation velocity σ^2 is zero in (2-8). Fault length, fault width, average dislocation velocity (average stress drop), and rupture velocity are the parameters which describe the deterministic part of the complex faulting

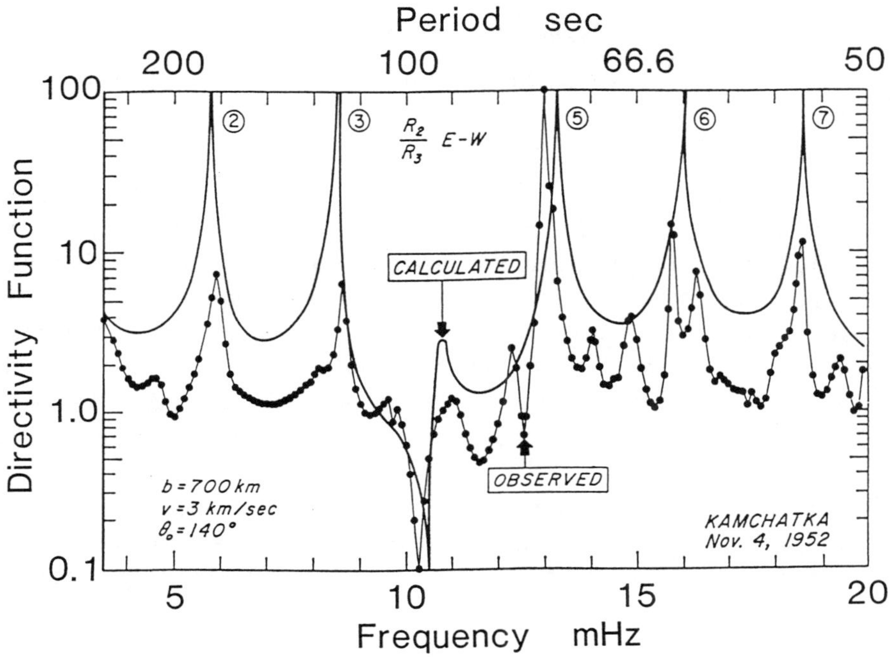

Figure 2 – 5. **Seismic directivity function at Pasadena, California for the 1952 Kamchatkan earthquake (solid circles). Fault length 700km, rupture velocity 3 km/sec, and rupture propagation direction of 140° clockwise from the north are obtained (Ben-Menahem and Toksöz, 1963).**

process. The corner frequency ω_c depends deterministically on the ratio of fault length and rupture velocity due to the finitely propagating fault.

Since T_0 is a rise time of dislocation motion, the term $T_0\bar{a}$ describes an average amount of dislocation on the fault. Therefore, $\frac{T_0\bar{a}}{W}$ is a measure of average strain release on the fault. Multiplying by rigidity μ, a relation between the average stress drop and the strain release is formally obtained as

$$\Delta\sigma_0 = \mu\frac{T_0\bar{a}}{W}. \qquad (2-14)$$

Stress drop in (2-14) is just a formal representation for the kinematical fault model. We use the dislocation model to describe the earthquake source in this text, although a crack model may also be used. When the stress drop of natural earthquakes is considered dynamically, it is necessary to apply the geometrical factor resulting from the fault shape to the kinematical expression in (2-14). The factor is a function of the ratio $\frac{W}{L}$ and is about 1.5 when $\frac{W}{L} = \frac{1}{2}$. The above equation could be also obtained from (2-8) and (2-9) with the definition of seismic moment

$$M_o = \Delta\sigma_0 W^2 L \ \ (= \mu L W T_0 \bar{a}). \qquad (2-15)$$

Mathematical theory of dislocations gives the equivalence of these two approaches for coherent rupture. For the complex faulting process, however, there is no theory that guarantees the validity of the equivalency in a general manner. All we have are that: the representation in (2-5) specifies the dislocation velocity on the heterogeneous fault, and that it describes the kinematics of the complex faulting process. Also, the average stress drop, defined in (2-14), formally specifies the average dynamical condition on the fault plane.

Similar to the variance dislocation velocity σ^2 in (2-5), the variance stress drop $< \Delta\sigma^2 >$ can be formally defined from the source spectrum in (2-8) to characterize the stochastic part of the complex faulting process :

$$< \Delta\sigma^2 > = \frac{2\mu^2\sigma^2 T_0}{W^2\lambda}. \qquad (2-16)$$

This can be derived from (2-8) in the limit $\omega \to 0$.

Because the variance stress drop is characterized by a random force with zero mean around the average stress drop $\Delta\sigma_0$, the fluctuation on localized fault patches does not contribute to the seismic moment. This gives an inequality relationship

$$\Delta\sigma_0^2 \gg < \Delta\sigma^2 > . \qquad (2-17)$$

The definition of λ in (2-6) is for the patch corner frequency introduced in order to describe the size effect of fault heterogeneities. The complex faulting process is thus formulated by the microscopic parameters of variance dislocation velocity σ^2, variance stress drop $< \Delta\sigma^2 >$, and patch corner frequency λ for the stochastic part of the complex faulting process, in addition to the macroscopic parameters for the deterministic part. Consequently, we need six independent parameters to describe the complex faulting process: L; W; $\bar{a}$; $\bar{v}$; λ and σ^2. Average stress drop, variance stress drop, and corner frequency are not independent parameters. From the natural properties of the earthquake faulting process, these six parameters may be reduced in number.

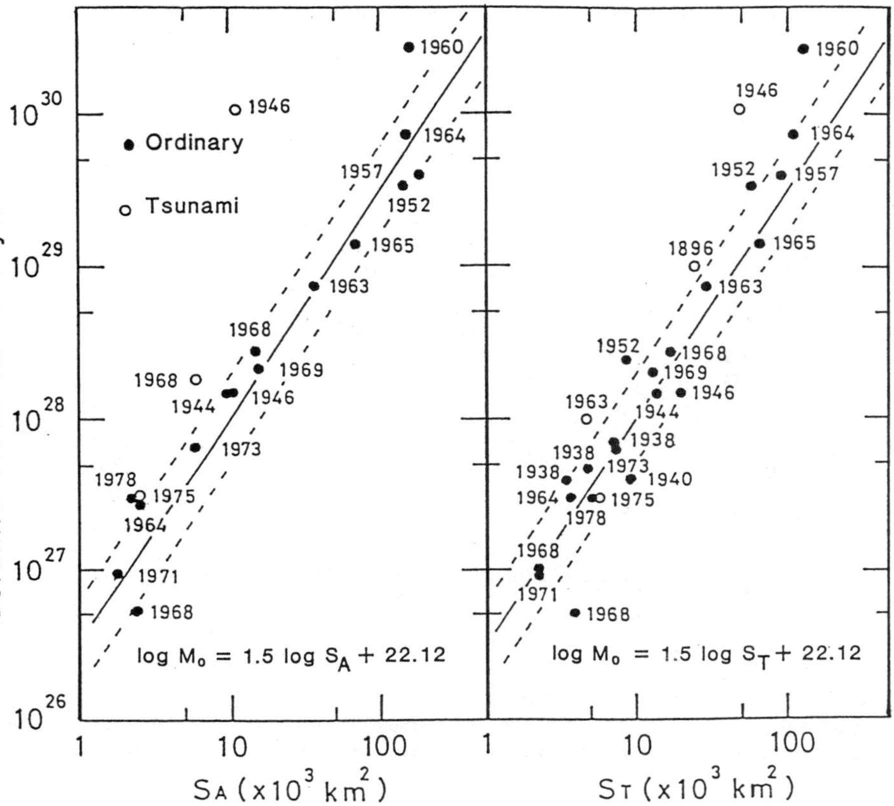

Figure 2 – 6. Aftershock area S_A and tsunami source area S_T versus seismic moment M_o of large shallow earthquakes. Numeral indicates the year of occurrence of each earthquake. Open and solid circles correspond to tsunami earthquakes (low-frequency earthquakes) and to ordinary earthquakes, respectively. Solid lines are plots of the empirical equations shown in the figure. Broken lines represent the uncertainty in M_o, S_A and S_T determinations, using a factor of 2 (Takemura and Koyama, 1983).

2.3. SCALING LAW OF EARTHQUAKE SOURCE PARAMETERS

2.3.1. *Geometrical Similarity*

Figure 2-6 examines an empirical relation between seismic moment M_o and source dimension S of large subduction-zone earthquakes. Two estimates of source dimensions are plotted, where S_A in Fig. 2-6(a) is from aftershock areas and S_T in Fig. 2-6(b) is from tsunami source areas. Solid lines in the figure indicate an empirical relation:

$$\log M_o = \frac{3}{2} \log S + 22.1, \qquad (2-18)$$

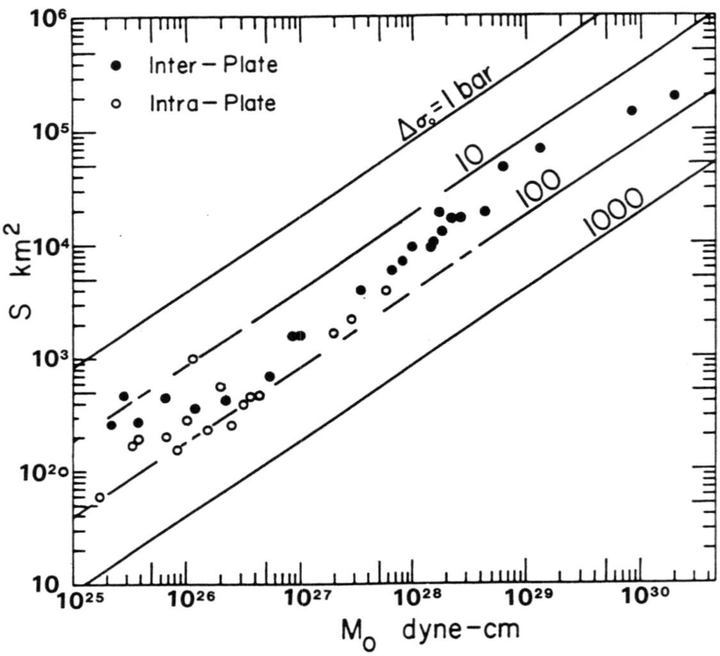

Figure 2 – 7. Relation between source dimension S and seismic moment M_o (Kanamori and Anderson, 1975). Straight lines assume a constant stress drop. Solid and open circles indicate the data from inter-plate and intra-plate earthquakes in the world, respectively.

where the unit of M_o and S is dyne·cm and km^2, respectively. Because S is reasonably understood to be a measure of LW, this relation suggests that the seismic moment is a function of only L and W. Taking a logarithm of the seismic moment in (2-15), the above equation is reduced to a relation

$$\frac{W}{L} = p \ (constant). \qquad (2-19)$$

This is a geometrical similarity of fault shapes irrespective of the earthquake source size.

2.3.2. *Dynamical Similarity*

The geometrical similarity suggests that seismic moment M_o is proportional to W^3. From (2-15), the constant coefficient is stress drop $\Delta\sigma_0$. This is a dynamical similarity of constant stress drop as

$$\Delta\sigma_0 = constant. \qquad (2-20)$$

This is found for natural earthquakes. Figure 2-7 shows a stress drop of about 30 bar for inter-plate earthquakes and about 80 bar for intra-plate earthquakes. The former events occur mostly along the shallow part of the subduction zones in the world and the latter occur within continental plates. This difference suggests a different seismic environment for different tectonic settings. Since the number of intra-plate earthquakes is relatively small, we focus on inter-plate earthquakes here.

2.3.3. *Kinematical Similarity*

By substituting the rise time T_0 in (2-4) into the expression for the stress drop in (2-14), the above dynamical similarity is shown to be equivalent to a kinematical similarity for the dislocation velocity

$$\bar{a} = constant. \qquad (2-21)$$

Another kinematical similarity has been independently obtained from the seismic Doppler effect of large shallow earthquakes

$$\bar{v} = constant. \qquad (2-22)$$

Rupture velocity, stress drop, and dislocation velocity are stable parameters, not only for earthquake faultings but also for rock failure in laboratory experiments.

The similarity laws discussed so far are derived by considering the macroscopic parameters of the complex faulting process. These similarity laws reduce the independent parameters to fault length (or fault width) in measuring the strength and size of the deterministic part of the complex faulting process. Although scaling stochastic source parameters is a very difficult problem, we will consider this next.

2.4. STOCHASTIC SIMILARITY OF COMPLEX FAULTING PROCESS

Short-period seismic waves from large earthquakes are investigated here in detail. Figure 2-8 shows short-period seismograms for the Tokachi-oki earthquake of May 16, 1968 and its aftershock of June 12, 1968 recorded at Berkeley, California, by the short-period Benioff seismometer. Its frequency characteristics is sharply peaked at about 1 Hz. Therefore apparent periods of P-waves on the seismograms are generally about 1 sec irrespective of earthquake source sizes. This provides us with information on the source excitation of seismic waves at about 1 sec, though waveforms are very complicated.

Wave trains following the direct P-wave arrivals are very different for the two earthquakes in Fig. 2-8. The wave train of the main shock takes about 80 sec to build-up to its maximum amplitude and almost the same time duration for decay. The hypocenters and point-source focal mechanisms of these two earthquakes are almost identical. Therefore the difference in

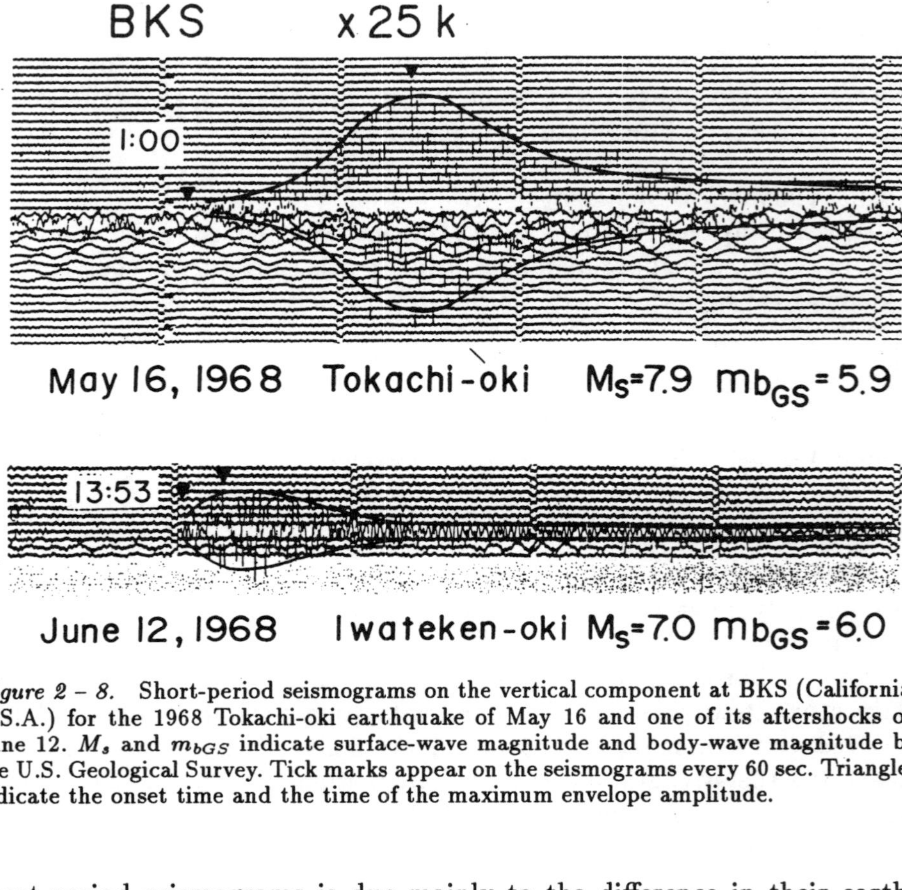

Figure 2 – 8. Short-period seismograms on the vertical component at BKS (California, U.S.A.) for the 1968 Tokachi-oki earthquake of May 16 and one of its aftershocks on June 12. M_s and m_{bGS} indicate surface-wave magnitude and body-wave magnitude by the U.S. Geological Survey. Tick marks appear on the seismograms every 60 sec. Triangles indicate the onset time and the time of the maximum envelope amplitude.

short-period seismograms is due mainly to the difference in their earthquake source processes. Body-wave magnitude m_b reported by the U. S. Geological Survey (USGS) is 5.9 for the main shock of May 16th and is 6.0 for the aftershock of June 12th. This indicates that m_b is not an appropriate measure for the short-period excitation of earthquake sources. This defect of the body-wave magnitude by the USGS results from its definition: m_b by the USGS is determined from the maximum amplitude of P-waves within the first few cycles and not from the maximum amplitude of P-wave trains.

2.4.1. *Seismic-waves with Random Phases*

Figure 2-9 shows seismograms of P-waves from the Alaskan earthquake of March 28, 1964, one of the largest events recorded by modern short-period instrument, and from other earthquakes with smaller magnitudes. These seismograms show that the larger the earthquake, the longer the rupture propagation time. The build-up time to the maximum of the envelope amplitudes in Fig. 2-9 is strongly dependent on the rupture propagation time

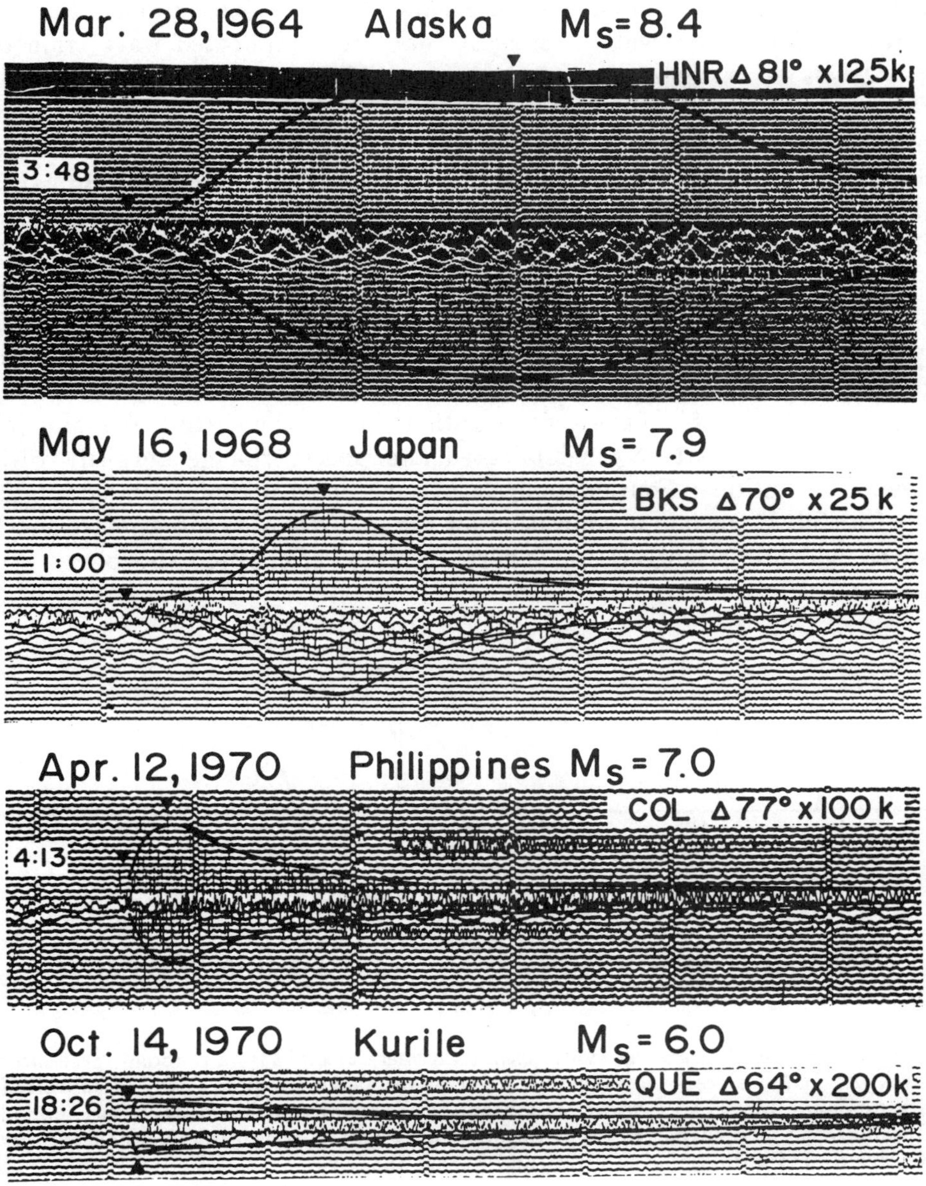

Figure 2 – 9. Short-period vertical component seismograms for earthquakes with various source sizes. The date and magnitude of each earthquake and the name of the seismic region are indicated. Three-letter station code, epicentral distance Δ in degree, and magnification are attached to each seismogram. It is seen that the longer the build-up time to the maximum amplitude, the larger the earthquake.

of each earthquake. The particle motion of wave trains estimated from three mutually perpendicular components is usually close to that of direct P-waves. These observations suggest that the complicated wave train of P-waves is generated at the source region. While, these wave trains may include reflected and converted phases (such as pP, sP, and SP), the relative contribution of these phases to the wave train amplitude does not change significantly from one earthquake to another, because all the earthquakes shown are shallow-focus earthquakes along subduction zones with similar thrust-type focal mechanism.

Short-period P-waves in Fig. 2-9 are characterized generally by (a) an apparent period of component waves within each wave train of about 1 sec irrespective of earthquake source size, (b) an envelope amplitude with gradual beginning and decay after the maximum, and for which build-up time depends on the rupture propagation time of the earthquake, and (c) waveforms are extremely complicated and phases change abruptly with time. Conventional Fourier analysis of such short-period seismograms is not fruitful. We need an innovation for the analysis of random-phase short waves, which incorporates the general characteristics listed above.

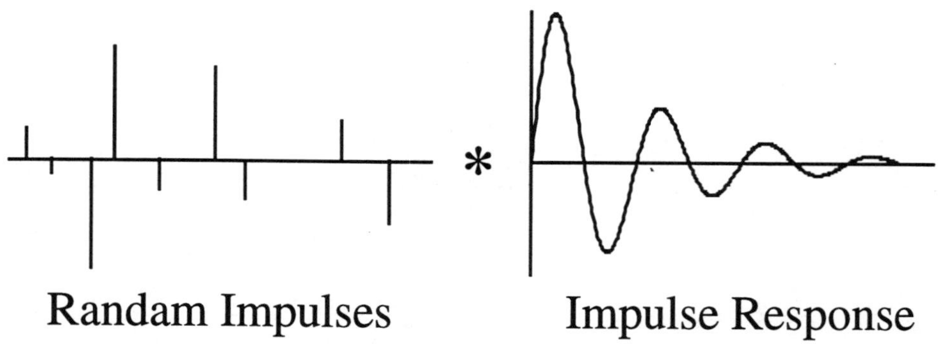

Figure 2 – 10. Modeling of the synthesis and the spectral analysis for random-phase short waves. An impulse response function of sine-waves tapered exponentially is convolved with random impulses to produce random-phase short waves with a characteristic period of oscillations. The output waves are windowed by an Gaussian error function with an effective time duration of τ (Koyama and Zheng, 1985).

2.4.2. *Synthetic Random-phase Short Waves*

Consider a component wavelet with finite duration and characteristic period. This wavelet represents the impulse response due to a rupture propagation on a particular fault patch. If this wavelet is activated by a random timing of shot noise process as in Fig. 2-10, the output response is random-

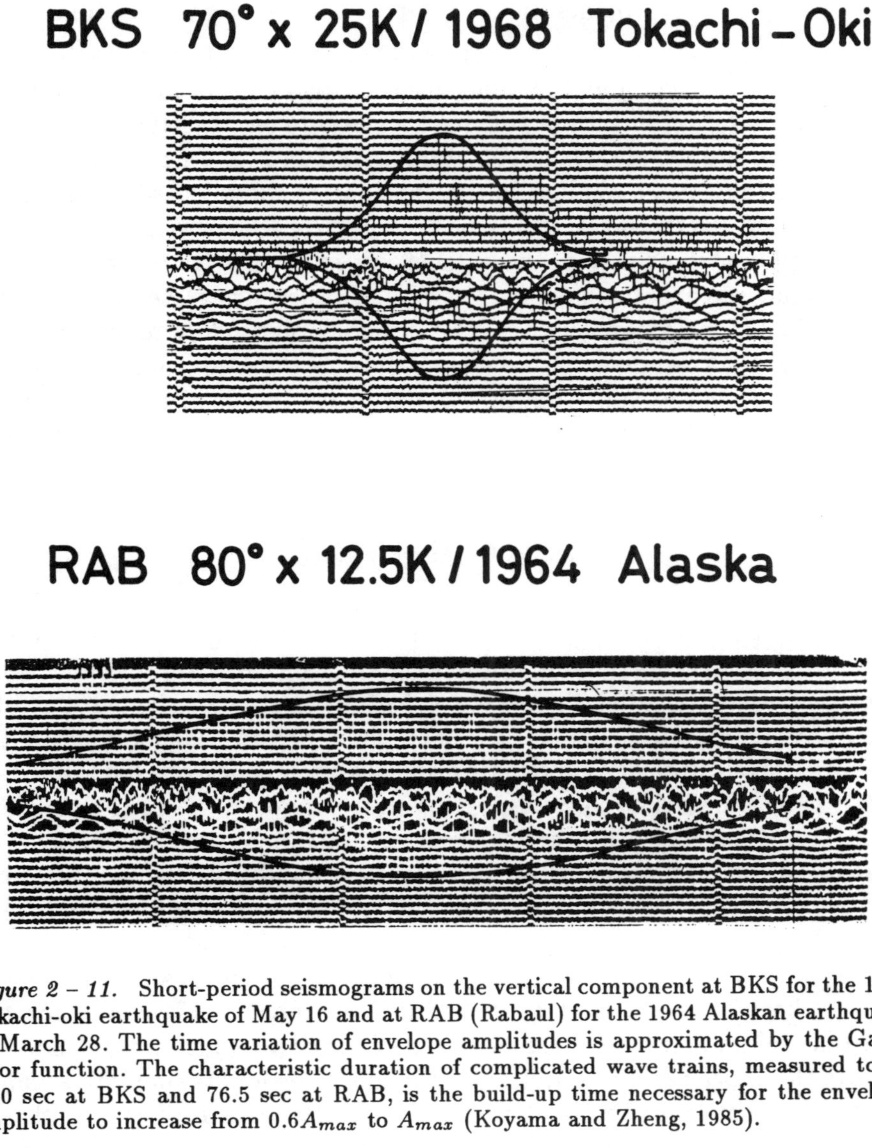

Figure 2 – 11. Short-period seismograms on the vertical component at BKS for the 1968 Tokachi-oki earthquake of May 16 and at RAB (Rabaul) for the 1964 Alaskan earthquake of March 28. The time variation of envelope amplitudes is approximated by the Gauss error function. The characteristic duration of complicated wave trains, measured to be 18.0 sec at BKS and 76.5 sec at RAB, is the build-up time necessary for the envelope amplitude to increase from $0.6A_{max}$ to A_{max} (Koyama and Zheng, 1985).

phase short waves. We investigate this as a model for seismic short-waves from the rupture process of random fault patches. The power spectrum for this model can be evaluated by Fourier analysis, since the component wavelet and random impulses are represented by convolution in the time domain.

Of many different types of functional forms, the Gauss error function

$\exp\left(-\frac{[t-t_{max}]^2}{2\tau^2}\right)$ provides the best fit to the temporal variation of the envelope amplitude of short-period seismograms. The time t starts at the onset of wave trains and the time interval to build-up to the maximum amplitude is t_{max}. The time constant τ determines the time duration of the envelope amplitude. Example seismograms in Fig. 2-11 show that this approximation can be applied to the envelope of complicated seismograms from different earthquakes with a wide variety of source sizes.

Multiplying the random-phase short waves by this Gaussian time window produces synthetic short-waves with the three characteristics (a), (b) and (c) in the previous section. This product of window function in the time domain corresponds to the convolution in the frequency domain. An approximate Fourier spectrum of random-phase short waves for this modeling is developed theoretically in Appendix B. The spectral peak which characterizes the apparent period of component short-waves is

$$\bar{Y}(\omega_b) \simeq 1.07\sqrt{\frac{2\pi\tau}{\omega_b}}\, A_{max}, \qquad (2-23)$$

where ω_b is the characteristic angular frequency of component short-waves and A_{max} is the maximum envelope amplitude. An approximate line-spectrum of complicated P-waves is calculated for each seismogram from three parameters A_{max}, τ and ω_b. The short-period seismograms in the present analysis are characterized by the peak frequency of about 1 sec, so that it is assumed that $\omega_b = 2\pi$ in this approximation. Since the envelope amplitude is expressed by the Gauss error function, τ can be measured as the time duration necessary for the envelope amplitude to increase from $0.6A_{max}$ to A_{max}. Verification of this spectral analysis is also given in Appendix B, by considering synthetic short-waves with random phases.

The line-spectrum thus estimated from each seismogram is corrected for the effect of the crustal magnification, assumed to be 2, and for the instrumental magnification. Equalization of wave travel distances is made to retrieve the short-period excitation at the source region, taking into account the geometrical spreading and the anelastic attenuation of the earth, by using the 68P-model and a short-period Q-structure (Table 2-1). Travel time divided by Q value along the seismic ray path (often designated t^*) is almost constant at about 1.4.

The spectrum is finally multiplied by $4\pi\rho\alpha^3$, giving a dimension of moment. Here, density ρ of 3.0 g/cm^3 and P-wave velocity α of 6.75 km/s are assumed. Spectral values thus evaluated at different stations are averaged for each earthquake and the average value is designated as short-period seismic excitation M_1 of the earthquake. Since we do not consider the contribution of reflected and converted phases included within P-wave trains,

TABLE 2 – 1. Geometrical spreading G and anelastic attenuation T/Q for P-waves from shallow earthquakes. Velocity structure of the 1968 P model (Herrin et al., 1968) and short-period Q structure by Veith and Clawson (1972) were used.

Delta (deg)	G (m^{-1})	T/Q (sec^{-1})	Delta	G	T/Q
30	7.21×10^{-8}	1.43	66	5.37	1.40
32	7.36	1.41	68	5.42	1.40
34	7.75	1.40	70	5.56	1.41
36	7.94	1.40	72	5.47	1.41
38	7.95	1.39	74	5.19	1.41
40	7.74	1.39	76	5.09	1.42
42	7.50	1.38	78	5.06	1.42
44	7.38	1.38	80	5.01	1.43
46	7.31	1.38	82	5.08	1.43
48	7.13	1.37	84	4.92	1.43
50	6.84	1.37	86	4.38	1.44
52	6.58	1.38	88	3.58	1.44
54	6.40	1.38	90	2.53	1.45
56	6.28	1.38	92	1.71	1.46
58	6.07	1.38	94	1.44	1.46
60	5.77	1.39	96	1.41	1.47
62	5.57	1.39	98	1.44	1.47
64	5.45	1.39	100	1.47	1.48

this estimate of short-period seismic excitation can not be compared to the seismic moment directly, but it is very important for relative characterizations of the earthquake strength in the short-period range.

2.4.3. *Stochastic Similarity*

To investigate the short-period seismic excitation M_1 for major earthquakes in the subduction zones, 50 events are selected. Surface-wave magnitude of all these earthquakes are larger than 7.0. Body-wave magnitude m_b^* has been redetermined using the maximum trace amplitudes of P-wave trains, so that it is not the same as m_b reported routinely by the USGS and the ISC. Seismic moments have been determined independently from long-period body-waves and from surface-waves. Table 2-2 summarizes the earthquakes analyzed. The long-period excitation of each earthquake can be estimated by seismic moment and surface-wave magnitude, and the short-period seismic excitation by M_1 and $m_b{}^*$.

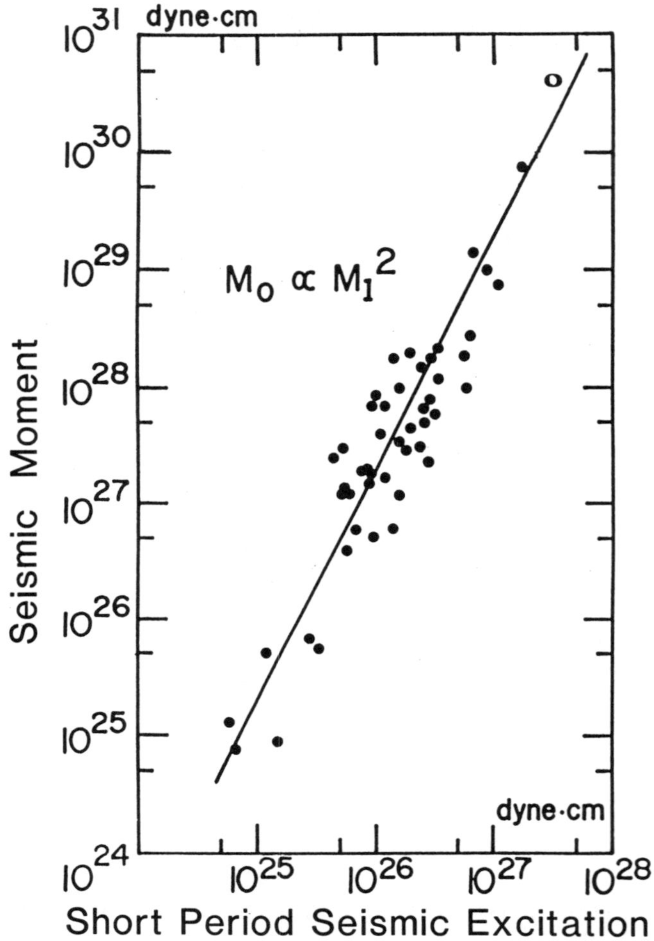

Figure 2 – 12. Seismic moment M_o and short-period seismic excitation M_1 of subduction zone earthquakes. Open circle indicates the data for the 1960 Chilean earthquake (seismograms provided by the courtesy of Heidi Houston). An empirical relation which systematically explains $M_o - M_1$ relation is illustrated by a solid line (Koyama and Zheng, 1985).

Figure 2-12 shows M_1 versus seismic moment M_o. There is no saturation of M_1 values over a wide range of seismic moments. Although some source models predict apparent saturation of short-period source excitation for large earthquakes, this result indicates that those models are inadequate.

The data scatter somewhat about a relation between M_1 and M_o which holds over a wide range of seismic moments. This empirical relation, obtained by a least squares fit in Fig. 2-12, is

$$M_o \propto {M_1}^2. \qquad (2-24)$$

While this relation does not constrain the absolute value of the short-period excitation, it is important for constraining the complex faulting process of earthquakes.

We consider the physical significance of this empirical relation. Two models are considered for the patch corner frequency λ of large subduction zone earthquakes. In the first case, λ is not a constant but relates to the fault dimension as

$$\lambda = c_p\, \omega_c, \qquad (2-25)$$

where ω_c is a corner frequency of the earthquake and c_p is a constant, about 7 to 10. This is designated as a scaling patch corner-frequency model (scaling P-model). Then the characteristic frequency $\omega_b (= 2\pi)$ for M_1 determination is generally much larger than λ of large and great earthquakes, because a typical value of ω_c for $M_S = 7.5$ earthquake is about $\left(\frac{L}{2\bar{v}}\right)^{-1} = 2{\times}3(\text{km/s})/100(\text{km})$. A high-frequency approximation of the source spectrum in (2-8) can be obtained at this frequency of ω_b, where both $\omega_b >> \frac{2}{T_0}$ and $\omega_b >> \frac{2\bar{v}}{L}$ are valid:

$$S_o \simeq \frac{2\bar{v}}{\omega_b}\left(\frac{2\lambda\sigma^2}{{\omega_b}^2 + \lambda^2}\right)^{1/2} W T_0^{1/2}. \qquad (2-26)$$

Then the theoretical representation of M_1 in the case of the scaling P-model is

$$M_1 \propto S_0 \simeq \frac{2\bar{v}}{{\omega_b}^2}(2\lambda\sigma^2)^{1/2} W {T_0}^{1/2}. \qquad (2-27)$$

Since $M_o \propto W^2 T_0$ from the similarity laws of the deterministic source parameters, the M_o-M_1 relation in (2-24) suggests an invariant parameter irrespective of the earthquake source sizes

$$\lambda\sigma^2 = constant. \qquad (2-28)$$

This is another similarity law for stochastic faulting parameters, and is designated the stochastic similarity of the complex faulting process.

In the second model, λ is a constant for large and great earthquakes (constant P-model). In this case, $\lambda \simeq 2\pi$, for reasons we will discuss later

TABLE 2 – 2. List of major earthquakes analyzed for short-period seismic excitation (Koyama and Zheng, 1985). Surface wave magnitude M_S has been quoted from Abe(1981). Body-wave magnitude m_b^* has been determined by Koyama and Zheng (1985). Seismic moments are quoted from Lay et al.(1982) and Purcaru and Berckhemer(1982).

Date	Seismic region	M_S	M_1 (dyne·cm)	m_b^*	M_o (dyne·cm)
May 22,1960	Chile	8.5	3.0×10^{27}	7.8	2.7×10^{30}
Oct.13,1963	Kurile	8.1	1.1×10^{27}	7.3	7.5×10^{28}
Oct.20,1963	Kurile	7.2	1.2×10^{26}	6.5	7.0×10^{27}
Mar.28,1964	Alaska	8.4	1.7×10^{27}	7.6	7.5×10^{29}
Feb. 4,1965	Rat Is.	8.2	6.7×10^{26}	7.1	1.4×10^{29}
Mar.16,1965	Japan		1.2×10^{25}	5.8	5.0×10^{25}
Mar.29,1965	Japan		3.3×10^{25}	6.1	5.5×10^{25}
Aug.23,1965	Mexico	7.6	7.6×10^{25}	6.7	1.9×10^{27}
Oct.17,1966	Peru	7.8	2.0×10^{26}	6.8	2.0×10^{28}
Dec.28,1966	N.Chile	7.7	2.0×10^{26}	6.9	4.5×10^{27}
Apr. 1,1968	Japan	7.6	9.2×10^{25}	6.7	1.8×10^{27}
May 16,1968	Japan	8.1	6.4×10^{26}	7.1	2.8×10^{28}
May 16,1968	Japan		2.8×10^{25}	6.2	6.7×10^{25}
May 20,1968	Kermadec	7.0	5.7×10^{25}	6.5	3.8×10^{26}
May 22,1968	Japan		5.9×10^{24}	5.5	1.3×10^{25}
June12,1968	Japan	7.3	9.5×10^{25}	6.5	5.1×10^{26}
Aug. 1,1968	Philippine	7.2	1.4×10^{26}	6.6	1.8×10^{28}
Nov.11,1968	Japan		6.5×10^{24}	5.7	7.5×10^{24}
Feb.28,1969	N.Atlantic	7.8	3.2×10^{26}	7.1	6.0×10^{27}
Aug.11,1969	Kurile	7.8	3.4×10^{26}	7.0	2.2×10^{28}
Nov.22,1969	Kamchatka	7.1	9.3×10^{25}	6.5	7.0×10^{27}
Apr.29,1970	Mexico	7.1	5.2×10^{25}	6.2	1.2×10^{27}
May 27,1970	Japan		1.5×10^{25}	5.9	9.0×10^{24}
May 31,1970	N.Peru	7.6	5.9×10^{26}	7.1	1.0×10^{28}
July26,1970	Japan		5.7×10^{25}	6.3	4.1×10^{26}
Jan.10,1971	New Guinea	7.9	2.9×10^{26}	7.0	8.0×10^{27}
July14,1971	Solomon	7.8	3.4×10^{26}	6.8	1.2×10^{28}
July26,1971	Solomon	7.7	3.0×10^{26}	6.7	1.8×10^{28}
Dec.15,1971	Kamchatka	7.5	1.6×10^{26}	6.7	1.0×10^{28}
July30,1972	Alaska	7.4	1.1×10^{26}	6.3	4.0×10^{27}
Jan.30,1973	Mexico	7.3	5.3×10^{25}	6.4	3.0×10^{27}
Feb.28,1973	Kurile	7.0	1.0×10^{26}	6.6	8.6×10^{27}
June17,1973	Japan	7.7	2.6×10^{26}	6.9	6.7×10^{27}
June24,1973	Japan	7.3	8.4×10^{25}	6.6	2.0×10^{27}
July 2,1974	Kermadec	7.0	1.6×10^{26}	6.7	1.2×10^{27}

(continued)

TABLE 2 – 2. Continue

Date	Seismic region	M_S	M_1 (dyne·cm)	$m_b{}^*$	M_o (dyne·cm)
July13,1974	Panama	7.1	1.4×10^{26}	6.7	6.1×10^{26}
Oct. 3,1974	Peru	7.6	2.5×10^{26}	6.8	1.5×10^{28}
May 26,1975	N.Atlantic	7.8	2.6×10^{26}	7.0	5.0×10^{27}
June10,1975	Kurile		4.4×10^{25}	6.1	2.5×10^{27}
July20,1975	Solomon	7.6	1.6×10^{26}	6.7	3.4×10^{27}
July20,1975	Solomon	7.5	5.9×10^{25}	6.2	1.2×10^{27}
Oct.31,1975	Philippine	7.4	2.8×10^{26}	7.0	2.3×10^{27}
Aug.16,1976	Mindanao	7.8	5.7×10^{26}	7.0	1.9×10^{28}
Aug.19,1977	Sumbaya	8.1	8.8×10^{26}	7.2	1.0×10^{29}
Oct.10,1977	Tonga	7.0	1.1×10^{27}	7.3	7.5×10^{27}
Mar.23,1978	Kurile	7.1	6.8×10^{25}	6.5	5.9×10^{26}
Mar.23,1978	Kurile	7.4	8.9×10^{25}	6.5	1.5×10^{27}
Mar.24,1978	Kurile	7.5	1.8×10^{26}	6.9	2.9×10^{27}
June12,1978	Japan	7.5	2.4×10^{26}	6.7	3.1×10^{27}
May 26,1983	Japan	7.8	4.9×10^{26}	7.0	5.1×10^{27}

again in §6. Assuming $\omega_b{}^2 + \lambda^2 = 2\omega_b{}^2$, since $\omega_b = 2\pi$, the high-frequency source spectrum in (2-26) is approximated as

$$S_0 \simeq \frac{2\bar{v}}{\omega_b{}^2}(\lambda\sigma^2)^{1/2}\, W T_0{}^{1/2}. \qquad (2-29)$$

The empirical relation between M_1 and M_o in (2-24) is reduced to the same similarity rule as in (2-28) for the stochastic source parameters in this case of the constant P-model.

This stochastic similarity is the clue for constraining the complex faulting process in the short-period range. We have studied the geometrical, dynamical and kinematical similarities to describe the deterministic part of the complex faulting process and here the stochastic similarity for the stochastic part. These similarity laws are essential for understand the scaling of earthquake sources. The values of the similarity parameters will be determined in §6 from observations.

CHAPTER 3

ACCELERATION SPECTRUM OF COMPLEX FAULTING PROCESS

3.1. ACCELERATION SPECTRUM OF HETEROGENEOUS FAULTING

Ground motion acceleration radiated from heterogeneous faulting can be derived from the displacement components of (2-1) to (2-3) by time differentiating twice. The acceleration spectrum for the complex faulting process is similarly derived from the displacement source spectrum in (2-9)

$$G_h(P,S) = \frac{M_o}{4\pi\rho r_0(\alpha^3,\beta^3)}|R_{\theta\phi}(P,S)|\ \omega^2|B_c(\omega)|. \qquad (3-1)$$

Due to the term of $\omega^2|B_c(\omega)|$, the acceleration spectrum shows a frequency dependence varying $\omega^2 - \omega^0 - \omega^{2-h} - \omega^0$ with increasing frequency. The corner frequency determines the change in the spectral envelope between ω^2 and ω^0, while the patch corner frequency is between ω^{2-h} and ω^0. Figure 3-1 shows examples of acceleration spectra from the complex faulting process. Hereafter, the acceleration spectrum in (3-1) is designated as that of the stochastic faulting model, in contrast with the spectrum evaluated from the self-similar fault patch, which is designated the self-similar faulting model.

Constant acceleration spectra have usually been found from strong motion records, these correspond to the spectral component ω^0 beyond the patch corner frequency in the above theoretical spectrum (3-1). It was shown in §2.1 that the spectral component in this frequency range is enhanced by the rupture propagation on random fault patches. In the intermediate frequency range from the corner to the patch corner frequencies, we observe an increase of spectral components in proportion to ω^{2-h} where h is about 1. The spectral component in this intermediate frequency range is due to the combination of the deterministic and stochastic parts of the complex faulting process. This spectral increase has been noted in acceleration records and also can be shown from the relations among different earthquake magnitudes.

Since strong motion acceleration is dominated by short-period S-waves, let us study the high-frequency approximation for S-wave acceleration in (3-1). Far-field approximation in (2-9) and high-frequency approximation here are always valid for short waves even in the near field, because the effective rupture radiating short waves is not the whole fault but small-scale fault patches.

Strong ground motion of large earthquakes always shows complicated waveforms. Apparently wave trains are composed of continuously arriving random-phase short waves. Figure 3-2 shows an example accelerogram for the Japan Sea earthquake of May 26, 1983, which illustrates the general property of strong ground motion. The random-phase rather than continuous phase ground motion is mainly due to the generation of short waves by

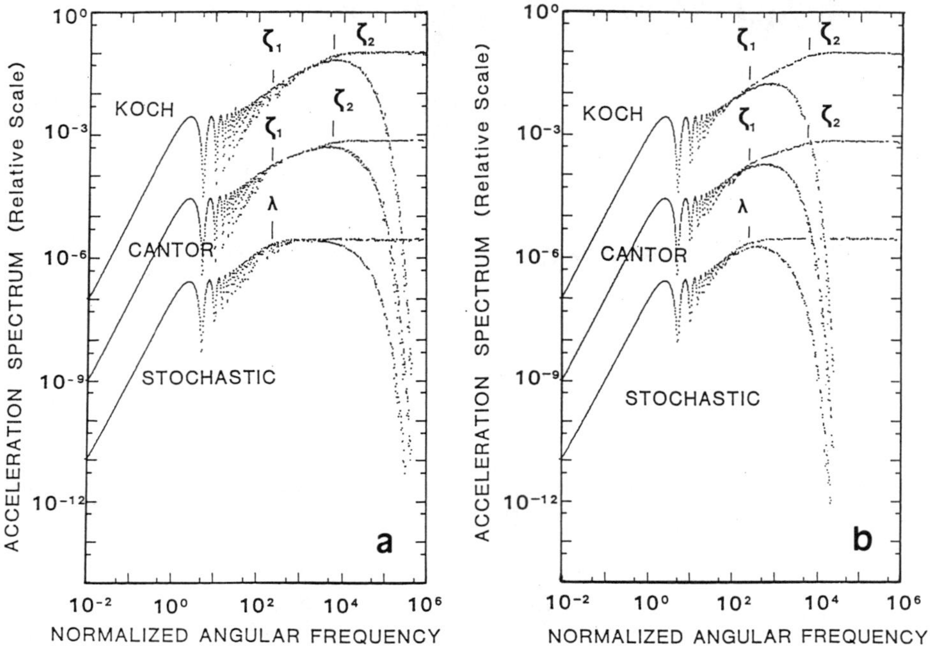

Figure 3 – 1. Acceleration spectrum from the complex faulting process. The stochastic faulting model and self-similar faulting models are considered, the latter are labeled as Cantor and Koch models (Koyama, 1994). The stochastic faulting model shows frequency dependence varying $\omega^2 - \omega^0 - \omega^{2-h} - \omega^0$, where h is about 1. The angular frequency here is normalized by W/β. The self-similar faulting models show a frequency dependence of $\omega^2 - \omega^0 - \omega^{2-h} - \omega^{\kappa/2} - \omega^0$, where κ is the fractal dimension of self-similar fault patches. Characteristic frequencies of ζ_1 and ζ_2 are due to the rupture processes of the largest and the smallest fault patches. In the high frequency range, all spectra show a constant spectral density. Spectral roll-off in the high frequency range is due to small attenuation in Fig. 3-1(a) and to large attenuation in Fig. 3-1(b): an empirical attenuation of the form $\exp(-\omega t^*/2)$ is assumed, where $t^* = 0.000039$ normalized by W/β in (a) and $t^* = 0.00069$ in (b), respectively.

the rupture processes of random fault patches. In addition to that, the focal mechanism of fault patches are not necessarily the same as that of the coherent rupture, but vary randomly. Such waves, after refraction, reflection and scattering, leave the source region. They consequently show a weak amplitude-dependence on the point-source focal mechanism of the deterministic source process. The total power of short-period acceleration at an observation r_0 can be derived after lengthy but straightforward calculations as an average:

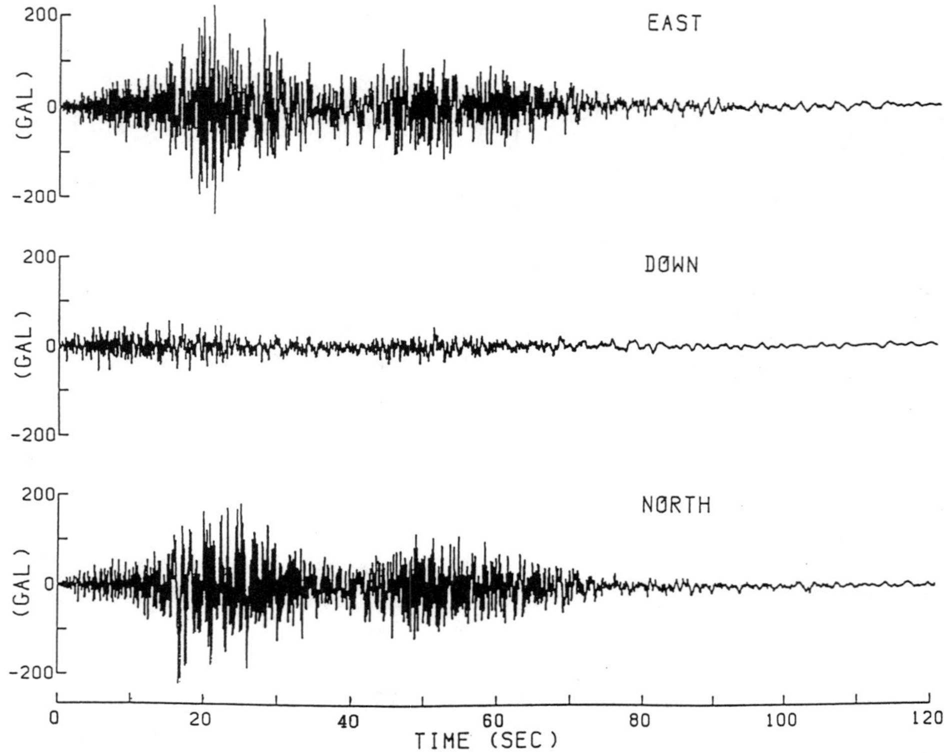

Figure 3 – 2. Strong motion acceleration at Akita (Port Harbor Research Institute), Japan in case of the 1983 Japan Sea earthquake of May 26. Three components of ground accelerations, east-west, down-up and north-south components, are shown. The ordinate of 200 gal is 20% of gravity. The phases of ground motion are not continuous as for sine waves but change abruptly from time to time. The double-peaked envelope of the seismograms is due to strong S-wave arrivals from two major subevents.

$$\begin{aligned} P_a(r_0) &= \frac{1}{4\pi}\int_0^{2\pi} d\phi \int_0^{\pi} \{P_t(u_\theta) + P_t(u_\phi)\} \sin\theta d\theta \\ &\simeq g(\chi)\Big(\frac{M_o}{4\pi\rho r_0 \beta^3}\Big)^2 \frac{<\Delta\sigma^2>}{\Delta{\sigma_0}^2} \frac{\lambda^2(\omega_m - \frac{\pi\lambda}{2})\bar{v}^2}{L^2}, \end{aligned} \quad (3-2)$$

$$g(\chi) = \frac{96 - 100\chi^2 + 16\chi^4}{3\pi\chi^4(1-\chi^2)} + \frac{8 - 3\chi^2}{\pi\chi^5} \ln\Big(\frac{1-\chi}{1+\chi}\Big)^2, \quad (3-3)$$

and χ is a velocity ratio of $\bar{v}/\beta$ and

$$P_t = \frac{1}{2\pi}\int_{-\omega_m}^{\omega_m} G_h(S)^2 d\omega, \quad (3-4)$$

where ω_m is an empirical cut-off frequency beyond which the spectral component does not contribute effectively to the actual strong ground motion. The *total power* in the above does not indicate its literal mean but rather represents the *total energy*. This usage is just conventional in applying Parseval's equality.

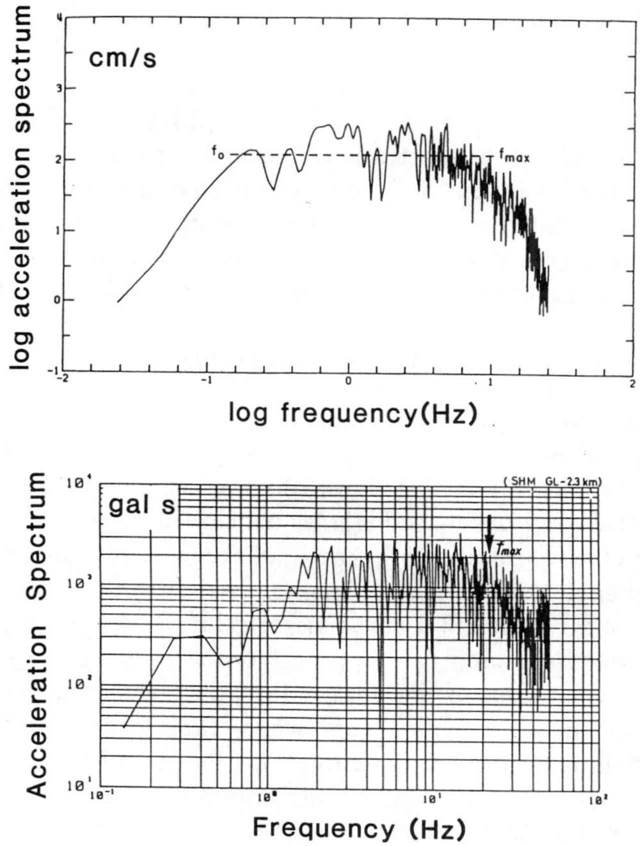

Figure 3 – 3. Fourier spectra of strong motion accelerations recorded at Pacoima dam (top) in case of the San Fernando earthquake of February 9, 1971, California (Hanks, 1981) and at Shimohsa in case of the 1987 East off Chiba earthquake, Japan (Kinoshita, 1988). Sharp spectral roll-off can be found in the frequency range higher than f_{max}.

Let us consider attenuation of high-frequency acceleration. An exponential decaying behavior of the type $\exp\left(-\frac{\omega t^*}{2}\right)$ is tentatively assumed, where t^* is the travel time of seismic waves divided by an average quality factor Q along the seismic ray path. We have seen in the previous chapter that t^* is a slowly varying function of distance and frequency. Acceleration spectra with this attenuation are also shown in Fig. 3-1. We can evaluate the total power in this case as

$$P_a'(r_0) = g(\chi)\Big(\frac{M_o}{4\pi\rho r_0\beta^3}\Big)^2 \frac{<\Delta\sigma^2>}{\Delta\sigma_0^2} \frac{\lambda^2\bar{v}^2}{L^2}\Big(\frac{1}{t^*} - \frac{\pi\lambda}{2}\Big). \qquad (3-5)$$

By comparing the total power in (3-2) with the above expression, we obtain

$$\omega_m = \frac{1}{t^*} \quad (\omega_m \gg \lambda). \qquad (3-6)$$

This shows that one possible origin for the empirical cut-off (angular) frequency ω_m of acceleration spectra is anelastic attenuation.

Physically significant in (3-2) and (3-5) is that the excitation of short-period accelerations is dependent on variance stress drop $<\Delta\sigma^2>$. This is essential to the complex faulting process, but few studies have included this important fluctuation of stress drop on the fault plane.

3.2. F_{MAX} AND SELF-SIMILAR FAULT PATCHES

3.2.1. *f_{max} due to Anelastic Attenuation*

An empirical cut-off frequency f_{max} has been noted on spectra of strong motion accelerations. Figure 3-3 shows observed spectra, where the abrupt decay of spectral components at high frequencies can be identified. The frequency which characterizes such behavior of acceleration spectra is denoted f_{max}. Its frequency range is much higher than that of the corner and patch corner frequencies. Figure 3-4 summarizes relations between seismic moments and corner frequency f_{corner}, patch corner frequency f_{patch}, and cut-off frequency f_{max}. Most of the data points, obtained from moderate-size earthquakes, are in a frequency range where three characteristic frequencies merge. It appears that f_{max} estimations for large earthquakes change slightly with respect to seismic moment, though the number of the data is too small to reach a definite conclusion. Let us consider the physical basis of the cut-off frequency f_{max} in more detail.

Three hypotheses have been proposed for the physics of the cut-off frequency f_{max} of acceleration spectra. In the first hypothesis, discussed in §3.1, the cut-off frequency ω_m is attributed to anelastic attenuation, where

$$\omega_m = 2\pi f_{max} = t^{*-1}. \qquad (3-7)$$

Attenuation of seismic energy at the source region is considered to give rise to f_{max} as well as that under the observation site. Because the acceleration spectra at very high frequencies are proportional to ω^0 irrespective of the earthquake source size and because the spectral decay is controlled by

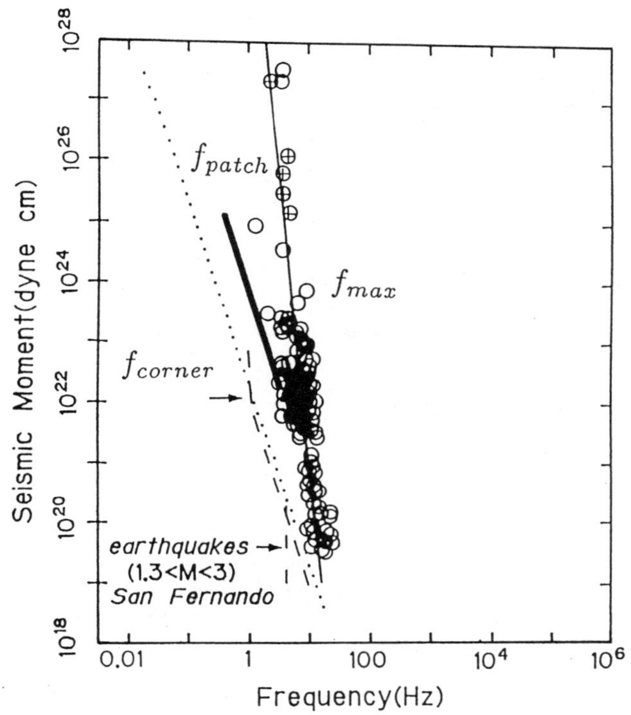

Figure 3 – 4. Corner frequency f_{corner}, patch corner frequency f_{patch}, and cut-off frequency f_{max} versus seismic moment. Dotted line represents a relation between f_{corner} and M_o derived from similarity rules among large earthquakes. Thick line represents a relation between f_{patch} and M_o (Koyama, 1985). f_{max} estimates (Aki, 1989) are plotted by circles against corresponding seismic moment values and thin line represents the general trend. Dashed lines represent relations between corner frequencies and seismic moments for earthquakes in and near San Fernando (Aki, 1989).

anelastic attenuation in this hypothesis, this hypothesis leads to a constant f_{max} irrespective of earthquake source size. As an example, we can find in Fig. 3-1 (b) where a strong attenuation controls the spectral behavior in high frequencies without any portion of a constant level of the acceleration spectrum.

3.2.2. f_{max} *due to Breakdown Zone*

The second hypothesis is based on the dynamics of crack-tip propagation. Consider a shear crack expanding at a velocity $\bar{v}$ (Fig. 3-5). A stress concentration appears at the front of the crack tip. The stress concentration is inversely proportional to the square root of distance from the crack tip. So that, the stress diverges to infinity at the crack tip. Such a stress singularity is physically unrealistic. A cohesive zone behind the crack tip has

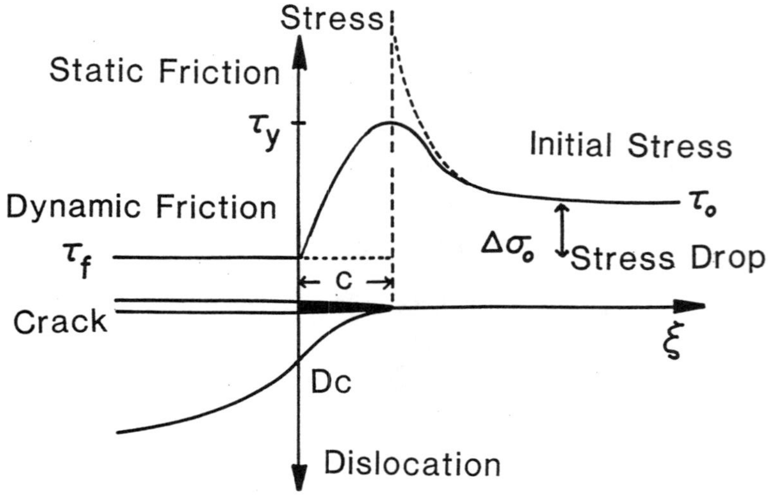

Figure 3 – 5. **Stress variation at a crack tip. Stress singularity at the crack tip is smeared out by a cohesive force which acts within the shaded zone with a length c. Parameter D_c is a critical dislocation, beyond which the stress drops to the dynamic friction level.**

been proposed to remove this stress singularity. It means physically that small-scale yielding takes place in this zone, and is represented by a cohesive force. The cohesive force produces anelastic deformation at the crack tip, removing the unrealistic stress singularity. This is the slip-weakening model described schematically in Fig. 3-5 and the zone where the cohesive force acts is called the breakdown zone.

When the stress concentration reaches τ_y in Fig. 3-5, slippage occurs to make new rupture surface. Cohesive force in the zone with length c reduces the amount of slip to a critical value D_c. Above D_c, the stress on the crack remains constant at τ_f, the dynamic friction. The existence of a cohesive force decelerates the propagation velocity of the rupture front, and also smooths the seismic impulse emitted at the instant of rupture onset. A time constant for the smoothing can be estimated from the breakdown of the cohesive zone. This smoothing effect has been proposed as the origin of f_{max}. This f_{max} hypothesis is presumed to be

$$f_{max} \simeq \frac{\bar{v}}{c}, \tag{3-8}$$

where c is the characteristic size of cohesive zones. The time constant $\frac{c}{\bar{v}}$ is said to be the rise time for the stress drop to increase from the static to the dynamic friction level at the crack tip.

The characteristic size of cohesive zones must be very small compared with the fault lengths and widths of large earthquakes. Once the rupture has propagated a sufficient distance, the size of cohesive zones depends only on the fault segment in the vicinity of the rupture front (the crack tip). This suggests that the time constant $\frac{c}{\bar{v}}$ is very small and that the step-function like variation of stress drop is for the earthquake faulting. As a consequence, it is likely that this hypothesis also suggests a constant cut-off frequency for the acceleration spectra of large earthquakes as did the first hypothesis in §3.2.1. These considerations lead to a weak dependence of f_{max} on earthquake source size.

3.2.3. f_{max} *due to Self-similar Fault Patches*

The third hypothesis to account for cut-off frequency f_{max} can be obtained from a scaling relation for the size-number distribution of random fault patches. Let us consider the minimum characteristic corner frequency ζ_1 resulting from the rupture of the largest fault patch; the characteristic corner frequency is related to the inverse of the rupture propagation time on the largest fault patch. Suppose that, compared with the largest fault patch, there are j-times more fault patches which have k-times higher characteristic corner frequency than ζ_1. Also, that there are j^2-times more fault patches specified by k^2-times higher characteristic corner frequency. This scaling is repeated by M-steps up to the largest characteristic frequency ζ_2. This procedure gives a self-similar set of random fault patches on the heterogeneous fault. Parameter j specifies the number density and parameter k specifies the size of the self-similar fault patches. The largest fault patch and the smallest fault patches provide limiting constraints on the scaling relation. Figure 3-6 schematically illustrates such self-similar fault patches on a heterogeneous fault plane.

The acceleration spectrum resulting from the self-similar faulting model can be expressed by the discrete sum of squared acceleration spectra of all the self-similar fault patches. This is true when there is no interaction among the rupture processes of the random fault patches:

$$G_f(P,S) = \frac{M_o|R_{\theta\phi}(P,S)|}{4\pi\rho r_0(\alpha^3,\beta^3)}\frac{\omega^2|\sin([k_c-\omega/\bar{v}]L/2)|}{|(k_c-\omega/\bar{v})L/2|}\left[\frac{\sin^2(\omega T_0/2)}{(\omega T_0/2)^2}\right.$$
$$\left.+\frac{2\zeta_1\sigma^2}{T_0\bar{a}^2}\frac{1-(j/k)}{1-(j/k)^{M+1}}\sum_{m=0}^{M}\frac{j^m}{\omega^2+(\zeta_1 k^m)^2}\right]^{1/2}, \quad (3-9)$$

where M calculates the level of scaling, and the largest characteristic corner frequency ζ_2 is $\zeta_1\ k^M$ for j^M pieces of the smallest fault patches.

Figure 3-1 shows normalized acceleration spectra of $G_f(S)$ for these self-similar faulting models, taking parameters of $j = 2,\ k = 3$ for the spectrum labeled by Cantor and $j = 4$, $k = 3$ for that by Koch. These acceleration spectra are more complicated than the stochastic faulting model described in (3-1) and shown in Fig. 3-1. The envelope spectrum of the self-similar faulting models is represented by a frequency-dependence of $\omega^2 - \omega^0 - \omega^{2-h} - \omega^{\kappa/2} - \omega^0$ with increasing frequency. The power coefficient κ relates to a fractal dimension of the self-similar fault patches

$$\kappa = \frac{\ln j}{\ln k}. \tag{3 - 10}$$

This spectral behavior reveals the existence of three corner frequencies; the common corner frequency arising from the rupture propagation on the entire faulting, two characteristic corner frequencies ζ_1 and ζ_2 resulting from the rupture process of the largest and the smallest fault patches. The second corner frequency ζ_1 corresponds to the patch corner frequency in the self-similar faulting models of Cantor and Koch.

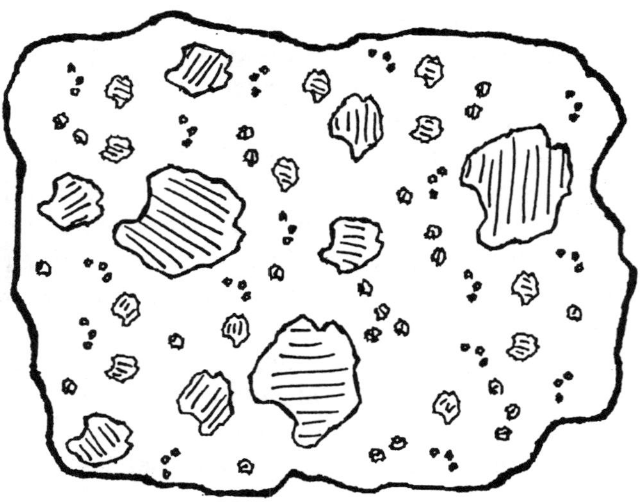

Figure 3 - 6. Self-similar fault patches on a heterogeneous fault are schematically illustrated. For fault patches is scaled down by k times in size, the number of fault patches increases by a factor j. The parameters j and k specify the fractal dimension of the scaling rule for the random fault patches.

Figures 3-1(a) and (b) also show acceleration spectra for small attenuation and large attenuation, respectively. High-frequency spectra are strongly distorted by the attenuation effect in Fig. 3-1(b). In Fig. 3-1(a), we find that the third corner frequency ζ_2 of the Cantor and the Koch spectra gives a sharp cut-off in the high-frequency acceleration spectrum. The sharp spectral decay after ζ_2 is much more drastic for the self-similar faulting models than for the stochastic faulting model. If we consider that this corner frequency corresponds to f_{max}, then the third hypothesis proposes a source-oriented f_{max} to explain the sharp spectral decay observed in acceleration spectra of large earthquakes. This hypothesis relates f_{max} to the rupture propagation on the smallest fault patches of the heterogeneous fault.

In summary, the third hypothesis proposes a source-oriented cut-off frequency f_{max}, and also describes the spectral behavior of high-frequency acceleration in terms of the size distribution of fault heterogeneities. Thus the complex faulting process of earthquakes introduces a new concept for the origin of f_{max}. Investigations have not determined which of these three hypotheses is correct, because direct measurements of such cut-off frequencies have been made for only a small number of earthquakes with limited frequency band observations. A scaling law for cut-off frequency, i.e. how the cut-off frequency changes with earthquake source size, is not yet completely determined.

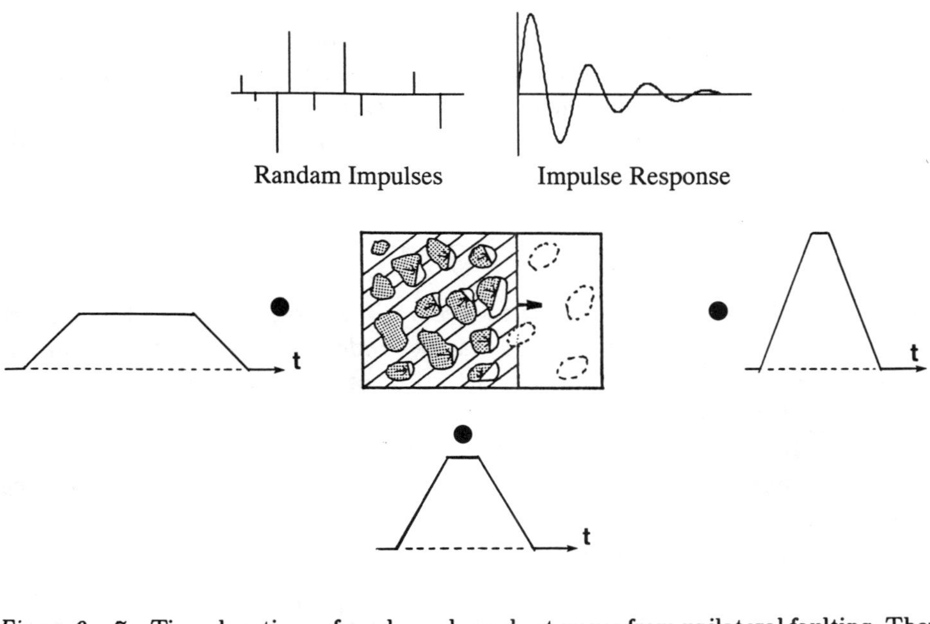

Figure 3 – 7. Time durations of random-phase short waves from unilateral faulting. They are illustrated by trapezoidal time functions in the rupture propagation, perpendicular to the rupture propagation and behind the rupture propagation directions. All the energy radiated from the heterogeneous fault is contained within corresponding time duration.

3.3. SHORT-PERIOD SEISMIC DIRECTIVITY

3.3.1. *Short-period Seismic Directivity of Unilateral Faulting*

Since energy must be conserved, random-phase short waves are square additive. Whereas phase coherency for continuous-phase long waves allows amplitude additive to conserve energy. This difference in short- and long-waves gives rise to a seismic directivity effect for short-period seismic waves different from the well-known seismic Doppler effect. Strong motion accelerations in Fig. 3-2 are complicated and characterized by random-phase. Generally speaking, strong motion accelerations from large shallow earthquakes at short epicentral distances are characterized by random-phase short waves. These waves are, therefore, square-additive and the squared sum is equal to the total power of acceleration spectrum in (3-2) and/or (3-5) through Parseval's equality. Accordingly, the root-mean-square amplitude of short-period acceleration a_{rms} at an observer can be defined by the squared sum of short waves and it is equal to

$$a_{rms}^2 = \frac{1}{T_d} P_a, \qquad (3-11)$$

where T_d is a time duration of a particular acceleration wave train and P_a is the total power in (3-2).

The effective time-durations of strong motion accelerations have been estimated for many large earthquakes from the cumulative plot of squared accelerations. These results show a close relation between the effective time-duration and the rupture propagation duration. Therefore, T_d is reasonably assumed to be the rupture propagation duration .

$$T_d = \frac{L}{\bar{v}}(1 - \frac{\bar{v}}{\beta} \cos\theta), \qquad (3-12)$$

where θ is the angle between the observation direction and the rupture propagation direction. This azimuthal dependence is illustrated schematically in Fig. 3-7, where the time duration of random-phase short waves changes as described in (3-12). Equation (3-12) is similar to the inverse of the corner frequency ω_β in (2-11) except for a factor of 2.

The azimuthal variation in (3-12) is due to a unilaterally propagating rupture on a finite fault. Consequently, the root-mean-square acceleration for this unilateral faulting is

$$a_{rms} = \left(\frac{P_a \bar{v}}{L}\right)^{1/2} \left(1 - \frac{\bar{v}}{\beta} \cos\theta\right)^{-1/2}, \qquad (3-13)$$

We see the effect of a finite rupture propagation on a_{rms} in (3-13). This is very important because the effect is not the same as the seismic Doppler effect on long waves.

The maximum acceleration may be estimated from the *rms* amplitude by the use of the statistical theory of extremes. A brief description of the theory is presented in Appendix C. The expectation of the maximum acceleration is expressed using the *rms* amplitude as

$$E[a_{max}] = \left\{ (2\ln N_a)^{1/2} + \frac{\gamma}{(2\ln N_a)^{1/2}} \right\} a_{rms}, \tag{3-14}$$

where N_a is the number of peaks and troughs within the acceleration wave train, and γ in (3-14) is Euler constant. A Rayleigh distribution, which has the probability density $p(x) = \left(\frac{2x}{a_{rms}^2}\right)\exp\left(-\frac{x^2}{a_{rms}^2}\right)$, has been assumed in this case for peak amplitudes. When peak and trough amplitudes are characterized by a Gaussian distribution, the most probable maximum acceleration is

$$M[a_{max}] = \left\{ \ln\left(\frac{N_a^2}{2\pi}\right) - \ln\ln\left(\frac{N_a^2}{2\pi}\right) \right\}^{1/2} a_{rms}. \tag{3-15}$$

Since the strong motion acceleration is usually characterized by a cut-off frequency of $\omega_m (= 2\pi f_{max})$, N_a is approximately expressed as the number of component waves of frequency ω_m within the time duration T_d:

$$N_a = \frac{T_d \omega_m}{\pi}, \tag{3-16}$$

where the factor of 2 has been considered in the above, which indicates two extreme values within one cycle of waves; peak and trough amplitudes. Equations (3-14) and (3-15) elucidate that the maximum acceleration is expressed in terms of *rms* acceleration, time duration of wave train and the characteristic frequency of component waves. The maximum acceleration is expected to increase as the time duration even though a_{rms} is the same. This indicates that the statistics of extreme values reveals that the chance of the largest amplitude with a small probability increases as the number of sampling (number of waves) N_a increases.

The number of peaks and troughs N_a changes as a function of T_d because of the seismic Doppler effect on T_d. However, the logarithm of N_a and the terms within braces in (3-14) and (3-15) changes very little with change of θ, provided that N_a is large. The dominant effect on the azimuthal variation of the maximum and *rms* amplitudes is, therefore, $(1 - \frac{\bar{v}}{\beta}\cos\theta)^{-1/2}$ for unilateral faulting. Because this azimuthal dependence on random-phase short waves results from a finite rupture propagation on the heterogeneous fault, this effect is designated the short-period seismic directivity effect. This contrasts to the seismic directivity effect for continuous-phase long

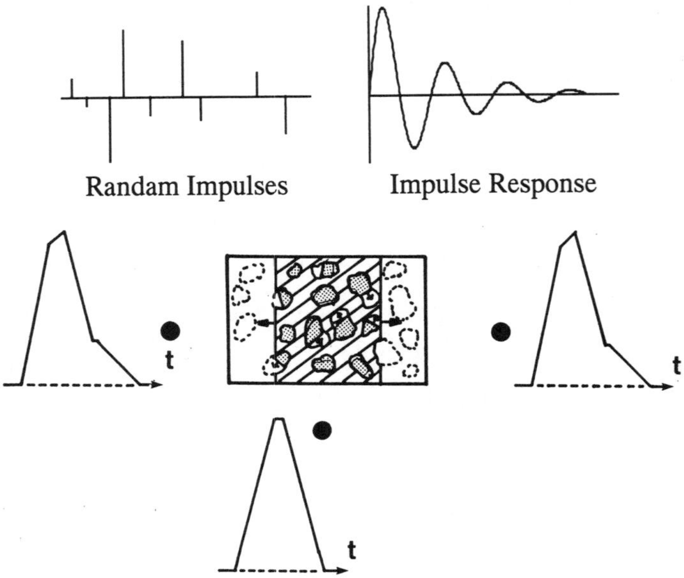

Figure 3 – 8. Time durations of random-phase short waves from bilateral faulting. They are represented by time functions composed of two trapezoidal functions and illustrated in the rupture propagation and perpendicular to the rupture propagation directions. All the energy radiated from the heterogeneous fault is contained within corresponding time duration.

waves, which has azimuthal dependence $(1 - \frac{\bar{v}}{\beta}\cos\theta)^{-1}$, referred to the seismic Doppler effect in §2.2.

3.3.2. *Short-period Seismic Directivity of Bilateral Faulting*

For a rupture which propagates in directions along both $\theta = 0$ and π (Fig. 3-8), we have a bilateral faulting. If the rupture extends at the same speed for a distance $\frac{L}{2}$ in both directions and that each fault segment has total power $\frac{P_a}{2}$, the *rms* acceleration can be approximated as

$$b_{rms} \simeq \left(\frac{2P_a\bar{v}}{L}\right)^{1/2}\left\{1 - (\frac{\bar{v}}{\beta}\cos\theta)^2\right\}^{-1/2}. \qquad (3-17)$$

Since the rupture duration of significant amplitude in this case is about one half that for unilateral faulting, the *rms* amplitude of bilateral faulting is larger than that of the unilateral faulting by a factor of $\sqrt{2}$ even though the net short-period excitation of the earthquakes are the same. This predicts larger acceleration for bilateral faulting than for unilateral faulting.

The maximum acceleration amplitude b_{max} of bilateral faulting can be derived similarly to that for unilateral faulting as

$$b_{max} = f(N_a)\, b_{rms}, \tag{3 - 18}$$

where $f(N_a)$ is found either in (3-14) or in (3-15). The dominant effect on the azimuthal variation of the maximum and rms amplitudes is, therefore, $\{1-(\frac{\bar{v}}{\beta}\cos\theta)^2\}^{-1/2}$ for bilateral faulting. This is symmetric along the fault strike, whereas that for unilateral faulting is asymmetric.

The difference of seismic directivity effects on long waves and on short waves appears in the power of the rupture propagation effect, that is, the inverse power for long waves and the square-root inverse for short waves. This difference in the seismic directivities results from the amplitude-additive and energy-additive natures of the different seismic waves. The short-period seismic directivity indicates that random phase reduces the azimuthal amplitude variation of short-waves compared with the case for long waves.

3.4. SHORT-PERIOD SEISMIC DIRECTIVITY ON STRONG MOTION ACCELERATIONS

There are obviously many difficulties in retrieving source information from short waves, one difficulty being the whole-path propagation effect due to anelasticity, geometric ray spreading and scattering in a three-dimensional heterogeneous medium. Amplitude magnification resulting from local shallow structures near the observation is another complicated factor. Teleseismic seismic waves leave the hypocentral region very quickly. Therefore, analyzing these waves as in §2.4 provides one technique for eliminating the effect of heterogeneous shallow structure near the source regions. Another approach is to assume aftershock records as an empirical Green's function to approximate the unknown whole-path propagation effect.

Although the treatment is not exhaustive, the effect of short-period seismic directivity on rms and maximum accelerations is investigated here. Figure 3-9 shows the ratio of rms strong motion accelerations of the 1968 Tokachi-oki earthquake on May 16th and of the largest aftershock on the same day. In the analysis we first define the time duration of strong motion accelerations as the time interval during which the cumulative integral of squared accelerations increases from 5 to 85% of its total. Root-mean square acceleration is then obtained from 80% of the total power of the acceleration in the two horizontal components. This is determined for the main shock and the largest aftershock independently at each station. Because the source regions and the focal mechanisms of these two earthquakes are known to be identical, the ratio of the main shock and the aftershock

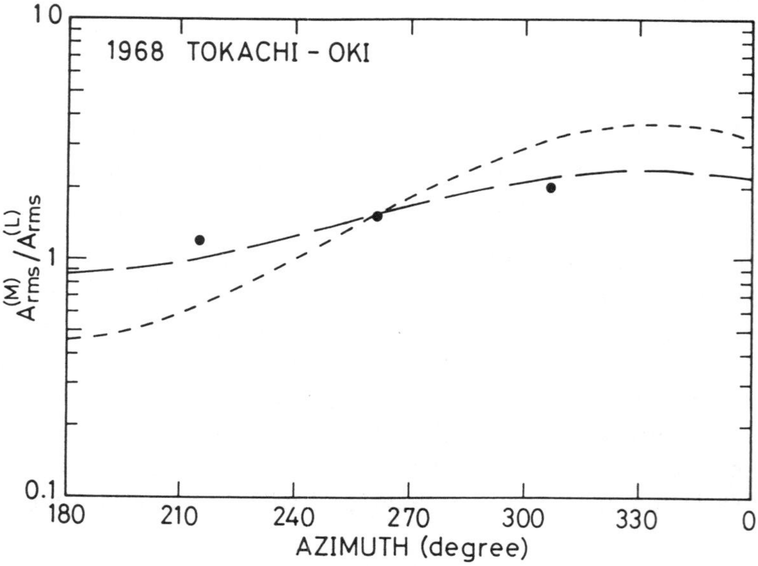

Figure 3 – 9. Ratio of root-mean square accelerations from the 1968 Tokachi-oki earthquake in Japan of May 16 and from its largest aftershock on the same day. The ratio is taken at each station to remove the uncertain correction for whole-path propagation effects. The long dashed curve indicates the short-period seismic directivity effect. The short dashed curve indicates the well-known seismic Doppler effect. Because of the distribution of earthquakes following the largest aftershock, the rupture propagation direction of the largest aftershock is assumed due south (Koyama and Izutani, 1990).

rms accelerations is, to some extent, free from effects due to the unknown whole-path attenuation.

We observe larger a_{rms} ratios in the northwest than in the southwest direction. This tendency is also found for the azimuthal variation of ratios of the maximum accelerations in Fig. 3-10. The rupture for the main shock of May 16th has been shown to be unilateral due northwest extending about 200 km. This provides a check on the short-period seismic directivity effect on the *rms* and maximum accelerations. The rupture propagation for the largest aftershock is assumed due south, on the basis of its epicenter location and aftershock distribution.

The long-dashed curves in Fig. 3-9 and Fig. 3-10 indicate the azimuthal variation of ratios calculated from the *rms* and maximum accelerations assuming the rupture propagation directions above. The expected variation from the seismic Doppler effect is indicated by short dashed curves and the short-period seismic directivity by long dashed curves in the figures. The azimuthal variation of short-period seismic directivity is small and matches

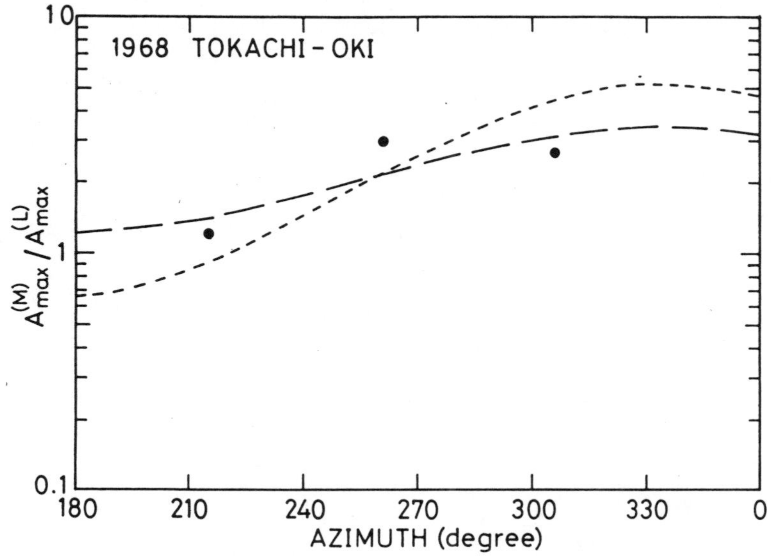

Figure 3 – 10. Ratio of the maximum accelerations from the 1968 Tokachi-oki earthquake and from its largest aftershock. For elaboration, see Fig. 3-9.

well with the observed variation in the figure. The variation of the seismic Doppler effect, on the other hand, larger than that of the observation for both the rms and maximum accelerations. This demonstrates that random-phase short waves reduce the azimuthal dependence of wave amplitudes.

3.5. SHORT-PERIOD SEISMIC DIRECTIVITY FROM ISOSEISMALS

Seismic intensity is an empirical scale for measuring the strength of strong ground motion near the earthquake epicenter. Its physical background is somewhat uncertain but seismic intensities have been obtained for many historical earthquakes and even currently it is very important for quantifying earthquake damage and to complement the sparse instrumental observations of strong ground motion. We can use the theoretical maximum acceleration in the empirical relation for modified Mercalli intensity. For unilateral faulting we have

$$I_M = 3 \log a_{max}(\theta) + 1.5, \qquad (3-19)$$

and for bilateral faulting

$$I_M = 3 \log b_{max}(\theta) + 1.5, \qquad (3-20)$$

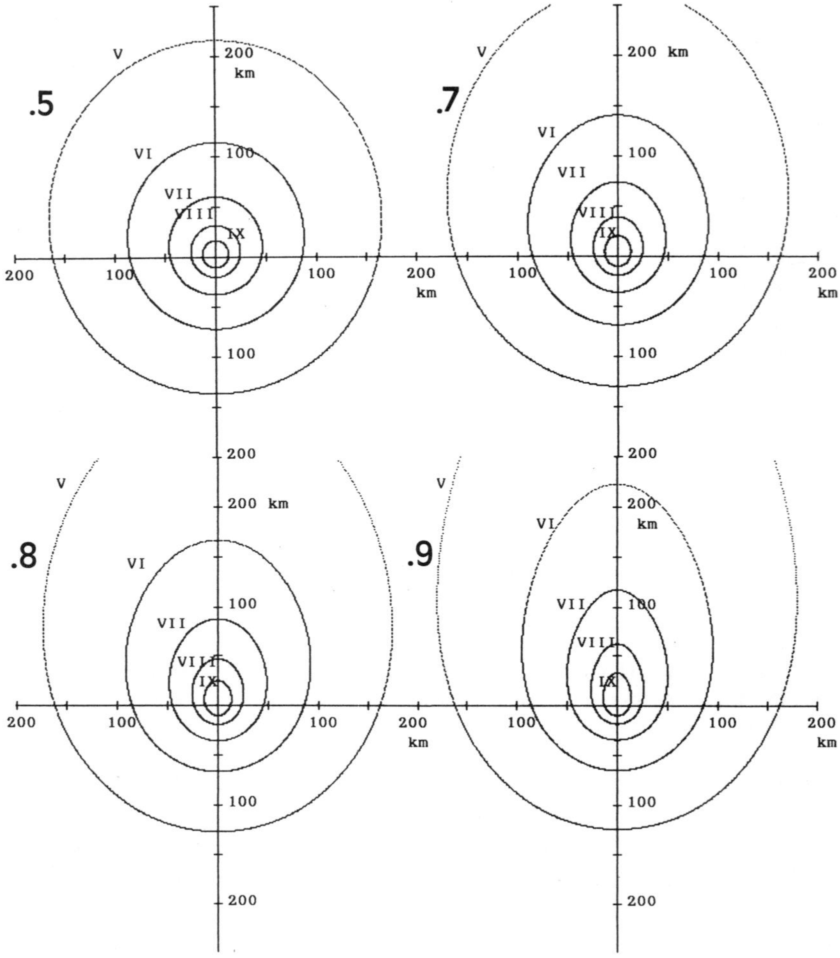

Figure 3 – 11. Isoseismals of unilateral faulting assuming rupture velocities, 50, 70, 80 and 90% of S-wave velocity. Model parameters are chosen to fit the seismic intensity map of the 1975 Haicheng earthquake, China using the modified Mercalli intensity. The ratio of major and minor axes of asymmetrical isoseismals changes with increasing rupture velocity. Egg shaped isoseismals are characteristic of the seismic intensity maps of unilateral faulting.

where a_{max} and b_{max} are the maximum accelerations (in gals) in the previous sections. Since a_{max} and b_{max} show azimuthal dependence due to short-period seismic directivity, the seismic intensity distribution should also reflect the faulting modes and rupture velocities.

Let us consider seismic intensity distributions, for rupture velocities of 50, 70, 80, and 90% of S-wave velocity. Figure 3-11 shows example seismic

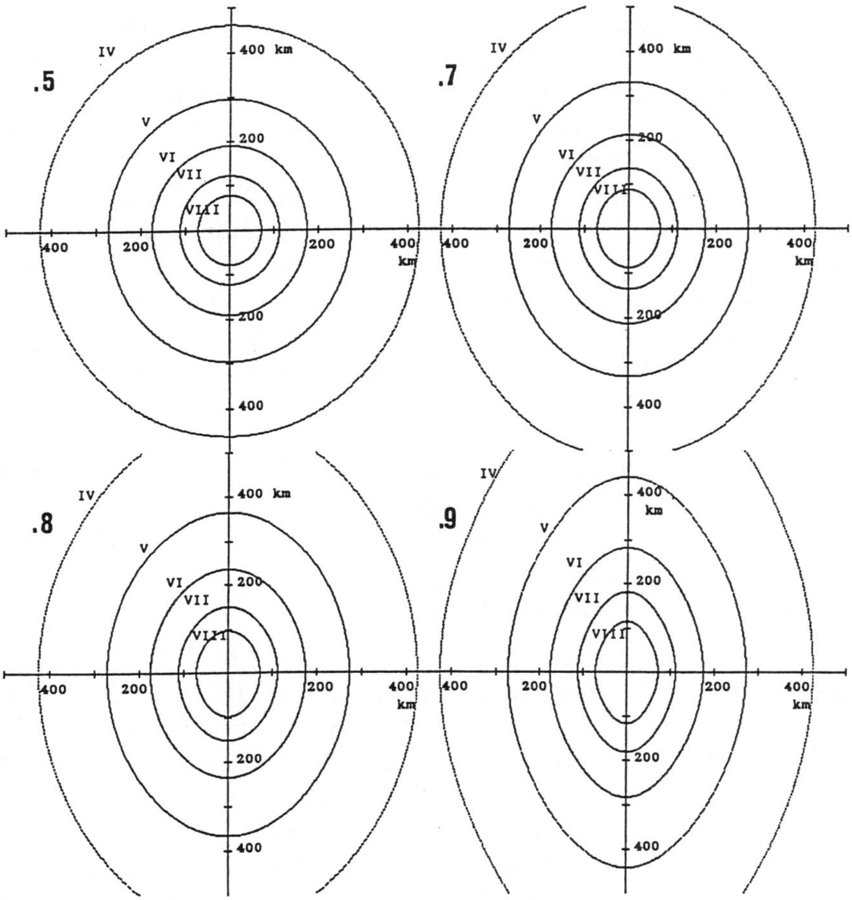

Figure 3 – 12. Isoseismals of bilateral faulting assuming rupture velocities, 50, 70, 80 and 90% of S-wave velocity. Model parameters are chosen to fit the seismic intensity map of the 1964 Niigata earthquake using modified Mercalli scale. The ratio of major and minor axes of elliptic isoseismals changes with increasing rupture velocity.

intensities calculated from (3-19). At a given epicentral distance seismic intensity for unilateral faulting is maximum in the rupture propagation direction, and minimum in the opposite direction. The distribution has an asymmetric egg shape. The difference between the maximum and minimum Mercalli intensities at the same epicentral distance is 0.7, 1.1, 1.4 and 1.9 for the above rupture velocities, respectively. This difference of about 1 in seismic intensity values is very large in terms of earthquake damage.

For a bilateral faulting the expected seismic intensity is maximum along the rupture propagation directions and minimum in the direction perpendicular to the fault trend. Thus the distribution has a symmetric ellipse shape. Figure 3-12 shows example seismic intensity distributions for (3-20). At a constant epicentral distance the difference in modified Mercalli intensities is 0.2, 0.4, 0.7, and 1.1 for the different rupture velocities, respectively. For high rupture velocity, the isoseismal has a very elongated elliptic shape, and becomes a circle-like for low rupture velocity. Therefore, the spread of isoseismals provides information on the rupture velocity as well as on the faulting mode for both unilateral and bilateral faultings. Although calculated values of intensity differences depend on the attenuation along the seismic ray path, the azimuthal differences above are very large, which encourages us to apply these theoretical seismic intensities to actual observations in order to study the earthquake faulting process.

3.5.1. *Isoseismals for Bilateral Faulting*

The 1964 Niigata earthquake in Japan and the 1976 Tangshan earthquake in China are discussed in detail. These earthquakes have been shown to have bilateral faulting by analyzing teleseismic body-waves. Field surveys have provided seismic intensity distributions covering wide azimuth angles (about 200°) and large epicentral distances.

Seismic Intensity of the 1964 Niigata Earthquake: The maximum seismic intensity of the 1964 Niigata earthquake was more than VII of modified Mercalli intensity in the epicentral area. Studies of long-period body-waves and of aftershock activity revealed that the quake had bilateral faulting extending about 50 km in both directions NNE and SSW.

The seismic intensity distribution for the Niigata earthquake is reproduced in Fig. 3-13. This distribution was obtained from the data of about 5500 questionnaires at every municipality. Since we have spatially dense information, the isoseismals of modified Mercalli intensity V, VI, and VII are clearly defined. Intensities at the same epicentral distance are large along the NNE-SSW direction and small in the ESE direction. Isoseismals look symmetric and show elliptic shapes. The major axis of the VII isoseismal is about 280 km and the minor axis about 200 km, a ratio about 1.4. The isoseismal of VI has a major axis about 500 km and a minor axis about 380 km. The ratio in this case is about 1.2.

Isoseismals in Fig. 3-13 are consistent with the theoretical pattern of seismic intensities shown in (3-20). We can calculate model isoseismals from (3-20) for the Niigata earthquake by assuming bilateral faulting with rupture velocity $\bar{v}$ as a parameter. The theoretical amplitude decay ${r_0}^{-1}$ in (3-20) is replaced by an empirical $\Delta^{-1.73}$, where Δ is epicentral distance.

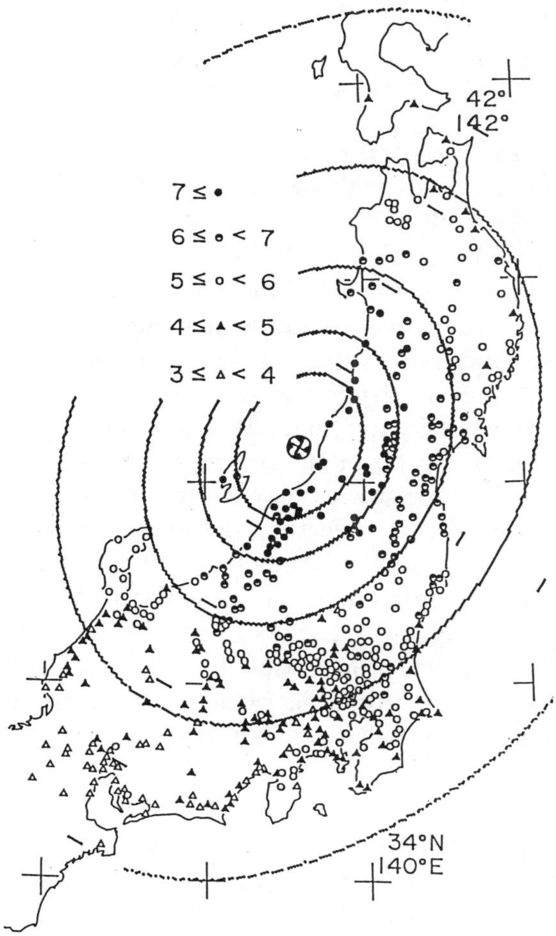

Figure 3 – 13. Seismic intensity map of the 1964 Niigata earthquake in Japan. Modified Mercalli intensity is evaluated at every municipality from data of about 5500 questionnaires (Kawasumi and Sato, 1968). Different symbols indicate corresponding seismic intensities. Cross within a circle marks the epicenter. Model isoseismals of bilateral faulting are shown by solid curves. They are calculated for a rupture velocity about 80% of S-wave velocity and fault strike N30°E. Note that the model isoseismals fit the observation in their symmetrical shape and spread.

This amplitude decay is the empirical standard correction for earthquake magnitude determination by the Japan Meteorological Agency (JMA). It is also known that this power decay of 1.73 applies to the empirical amplitude decay of the maximum accelerations with epicentral distance in Japan. The model isoseismal is compared with the observations in Fig. 3-13, the rupture velocity and faulting direction. The best fit to determine model isoseismal indicates that the rupture velocity of the Niigata earthquake was about 80% of S-wave velocity.

Isoseismals of bilateral faulting corresponding to the above rupture velocity are illustrated in Fig. 3-13. We have assumed a reference seismic intensity of VII at an epicentral distance of 100 km perpendicular to the fault trend. The model isoseismals of VI and VII match the observations in the direction of the fault and also in the perpendicular direction. Some inconsistencies are found in Fig. 3-13; for example, the observed seismic intensity of V is limited in the SSW direction compared with the NNE direction. There are many reasons for such minor discrepancies: (1) the Niigata earthquake bilateral faulting was larger in the NNE direction; (2) amplitudes of short-period seismic waves decay much faster in the SSW direction than in the NNE direction; (3) since intensity observations in the SSW direction were mostly in the central mountainous area of Japan, amplification of seismic short-waves may be systematically smaller in this region. A combination of these effects is the likely explanation of the minor inconsistencies in Fig. 3-13. Although not all the observations can be explained by the model isoseismals, the essential features of the seismic intensity map for the Niigata earthquake are reproduced by the bilateral faulting model. Thus rupture propagation direction and rupture velocity of the faulting process can be estimated from the isoseismal distribution on the basis of the short-period seismic directivity.

Seismic Intensity of the 1976 Tangshan Earthquake: Long-period surface-waves showed that the Tangshan earthquake of 1976 was a strike-slip bilateral fault on a vertical plane in the N220°E direction. The fault extended from the epicenter to both ends by as much as 70 km. The seismic intensity scale in China is based on modified Mercalli intensity scale adjusted for Chinese conditions. Well-trained specialists survey the earthquake source area and its vicinity, investigating the damage in structures and buildings, cracked plaster and fallen chimneys, ground deformations and landslides, damage on river banks and railways, and also human impressions. They then determine the seismic intensity at each place.

Maximum Mercalli intensity of XI occurred in the epicentral area of the Tangshan earthquake, where all buildings were completely destroyed. Figure 3-14 shows isoseismals of the Tangshan earthquake. Similar elliptic isoseismals surround the epicentral area except for some smaller areas of anomalously large seismic intensities such as Tianjin city. A similar elliptic isoseismal is obtained even for intensity XI at the very contour of the epicentral region.

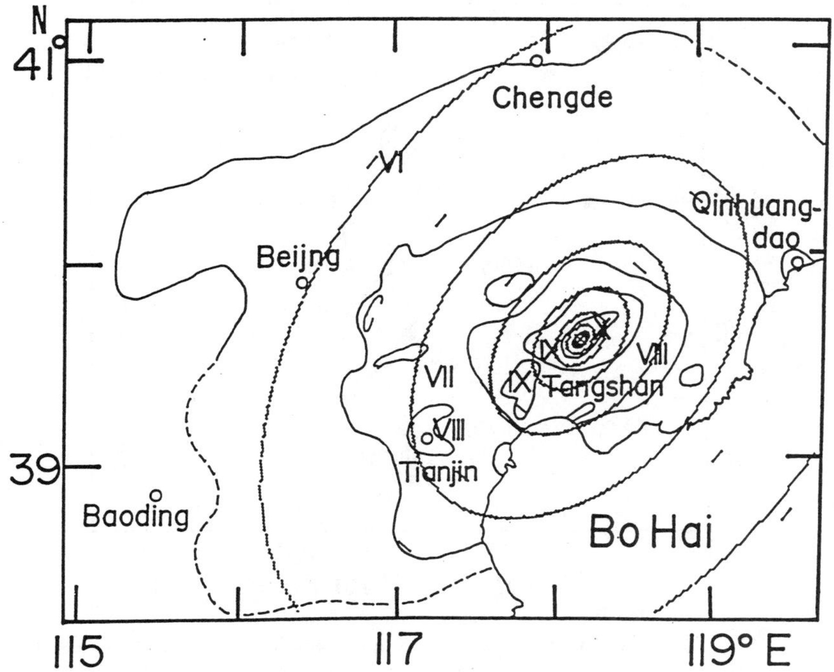

Figure 3 – 14. Isoseismal curves of the 1976 Tangshan earthquake in China (Mei, 1982). Modified Mercalli intensity, adjusted for Chinese conditions (Hsieh, 1957), has been estimated from the damage in structures and buildings. Model isoseismals of bilateral faulting are calculated assuming that the rupture velocity is 80% of S-wave velocity and that the fault strike is N220°E. Model isoseismals with similar shapes agree with the observations except for some regions of abnormal seismic intensities.

Figure 3-15 shows the length of major and minor axes of isoseismals of the 1976 Tangshan earthquake and of the 1976 Szechwan earthquake, Qionglai mountainous Province in China. These two earthquakes occurred in different tectonic regions. The seismic intensity at large epicentral distances decreases approximately as

$$I_M \propto -3.6 \log \Delta, \qquad (3-21)$$

where Δ is the epicentral distance. The power-law decay of the maximum acceleration from this formula is consistent with that for amplitude decay of L_g waves in China (Cheng and Nuttli, 1984).

Since the attenuation along seismic ray path is characterized by a power decay of 3.6 in China, smaller than the value of 3×1.73 in Japan, the ratio of the major and minor axes of isoseismals becomes a little larger

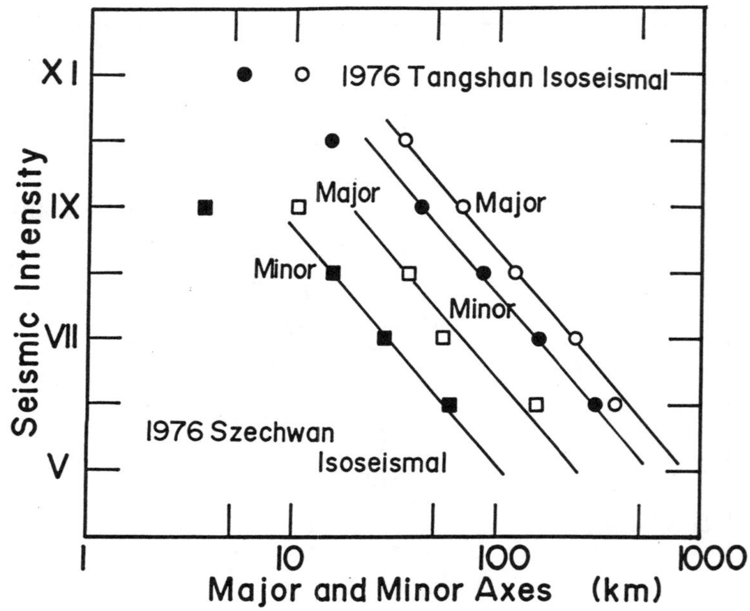

Figure 3 – 15. The length of major and minor axes of each isoseismal for the 1976 Tangshan earthquake in the great Huobei basin and the 1976 Szechwan earthquake in the Qionglai mountainous region in China. Solid lines indicate the variation of seismic intensities with epicentral distance. Although the two earthquakes occurred in different seismic regions, the variations are similar (Koyama and Zheng, 1991).

than that for the previous result in Japan. Figures 3-14 and 3-15 suggest that the observed ratios are about 1.46 for intensities of VI to XI of the Tangshan earthquake. We know that this ratio changes as a function of the rupture velocity. The observation favors a rupture velocity for the Tangshan earthquake is about 80% of S-wave velocity. Figure 3-14 also shows the model isoseismals from (3-20), on the basis of reference intensity VII at 75 km away from the fault in the direction perpendicular to the fault trend of N220°E.

The model isoseismals of VIII, IX and X agree well with the observations in terms of their general patterns of spreading and their areas. Therefore, the isoseismals of the Tangshan earthquake are well explained by this bilateral faulting. However, the observed isoseismals of VI and VII in Fig. 3-14 show large spreading in the ENE-WSW direction. It has been suggested that a small reverse faulting accompanied the Tangshan earthquake. This complication to the faulting process would have affected the isoseismals. The isoseismals spread over the great Huobei basin, a highly populated

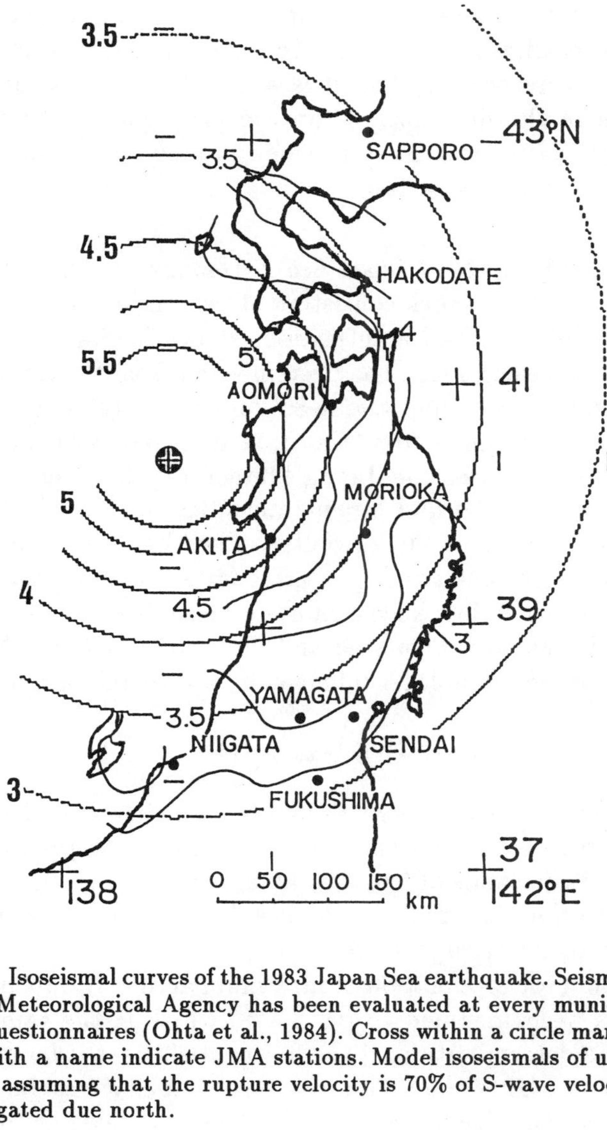

Figure 3 – 16. Isoseismal curves of the 1983 Japan Sea earthquake. Seismic intensity scale by the Japan Meteorological Agency has been evaluated at every municipality from the data of 8500 questionnaires (Ohta et al., 1984). Cross within a circle marks the epicenter. Solid circles with a name indicate JMA stations. Model isoseismals of unilateral faulting are calculated assuming that the rupture velocity is 70% of S-wave velocity and that the rupture propagated due north.

area in contrast to the surrounding mountainous areas. This could also distort the isoseismals but does not materially effect the basic seismic intensity distribution.

Thus the seismic intensity distribution of the Tangshan earthquake is well explained by bilateral faulting. We are able to retrieve reliable quantitative information on the strike direction and the rupture velocity from the seismic intensity observation.

3.5.2. *Isoseismals from Unilateral Faulting*

Seismic intensity distribution due to unilateral faulting is asymmetric and shows higher intensities in direction of rupture propagation. Higher rupture velocities lead to flatter egg-shaped isoseismals and to greater spread of the isoseismals in the direction of rupture propagation. This spread and asymmetry provide critical clues for determining the faulting mode and the rupture propagation direction.

Seismic Intensity of the 1983 Japan Sea Earthquake: The 1983 Japan Sea earthquake is a multiple shock consisting of two major subevents. This was seen from the double-peaked accelerations shown in Fig. 3-2. The rupture propagation of this earthquake has been inferred from the aftershock activity to be unilateral, extending about 80 km due north. Figure 3-16 shows seismic intensities obtained from the data of about 8500 questionnaires. This seismic intensity uses the Japan Meteorological Agency(JMA) scale as precise as 0.1 unit. We find larger intensities to the north and smaller intensities to the south of the epicenter. The three dimensional velocity structure in the North-eastern Japan arc does not explain such a distribution; we are required to invoke the short-period seismic directivity effect.

Model isoseismals are calculated and illustrated in Fig. 3-16, using the empirical relation between JMA intensity I_J and maximum acceleration

$$I_J = 2 \log a_{max} + 1.2. \qquad (3-22)$$

To obtain agreement with observations we use a reference JMA intensity of 4.0 at 200 km from the epicenter and a rupture velocity 70% of S-wave velocity. Model isoseismals of 5.0, 4.5, 4.0 agree well with the observations. Higher rupture velocities result in the isoseismal spacing seeing too large, and conversely for lower rupture velocities.

The correction for attenuation of maximum acceleration with epicentral distance Δ has been assumed to be $\Delta^{-1.73}$ which is the same as for the JMA magnitude determination. If the power coefficient is larger than 1.73, a rupture velocity higher than 80% of S-wave velocity may be permissible. Knowledge of the amplitude decay with seismic ray path is important for obtaining a precise estimate of the earthquake faulting mode and rupture velocity.

Seismic Intensity of the 1975 Haicheng Earthquake: The 1975 Haicheng earthquake in China had a left-lateral strike slip on a vertical fault plane striking N228°E. Figure 3-17 shows the seismic intensity distribution from the compiled data. Using (3-19), unilateral faulting gives the ratio of major to minor axes of model isoseismals as about 1.1, 1.2, 1.3 and 1.5 for the

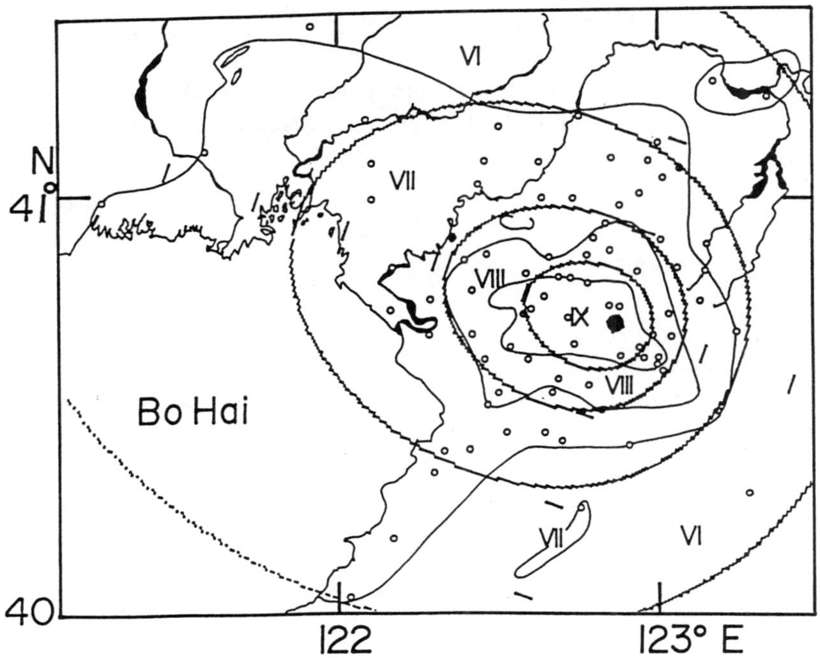

Figure 3 – 17. Isoseismal curves of the 1975 Haicheng earthquake in China (Jiang, 1978). The epicenter is shown by a solid circle. Small open circles indicate local municipalities. Model isoseismals of unilateral faulting are shown for rupture velocity 80% of S-wave velocity with faulting extending due N288°E. Larger major axes than those of the model isoseismals may suggest some bilateral component for the earthquake.

rupture velocities of 50, 70, 80 and 90% of S-wave velocity, respectively. Spacing of isoseismals in Fig. 3-17 and the observed ratio of axes favor a rupture velocity of 80% of S-wave velocity. Model isoseismals are illustrated in Fig. 3-17, where intensity VII at 45 km perpendicular to the fault is taken as a reference.

Matching of model isoseismals to the observations indicates that the Haicheng earthquake had unilateral faulting to the south-west. Since the epicenter of the Haicheng earthquake is located in the middle part of the aftershock area, it apparently had some bilateral component. This may be why the observed ratios of the major to the minor axes are a little larger than for the theoretical model.

3.6. EXCITATION OF SHORT-PERIOD ACCELERATIONS

Directivity effects on strong motion accelerations have been studied for earthquakes in California; these are often interpreted in terms of the seis-

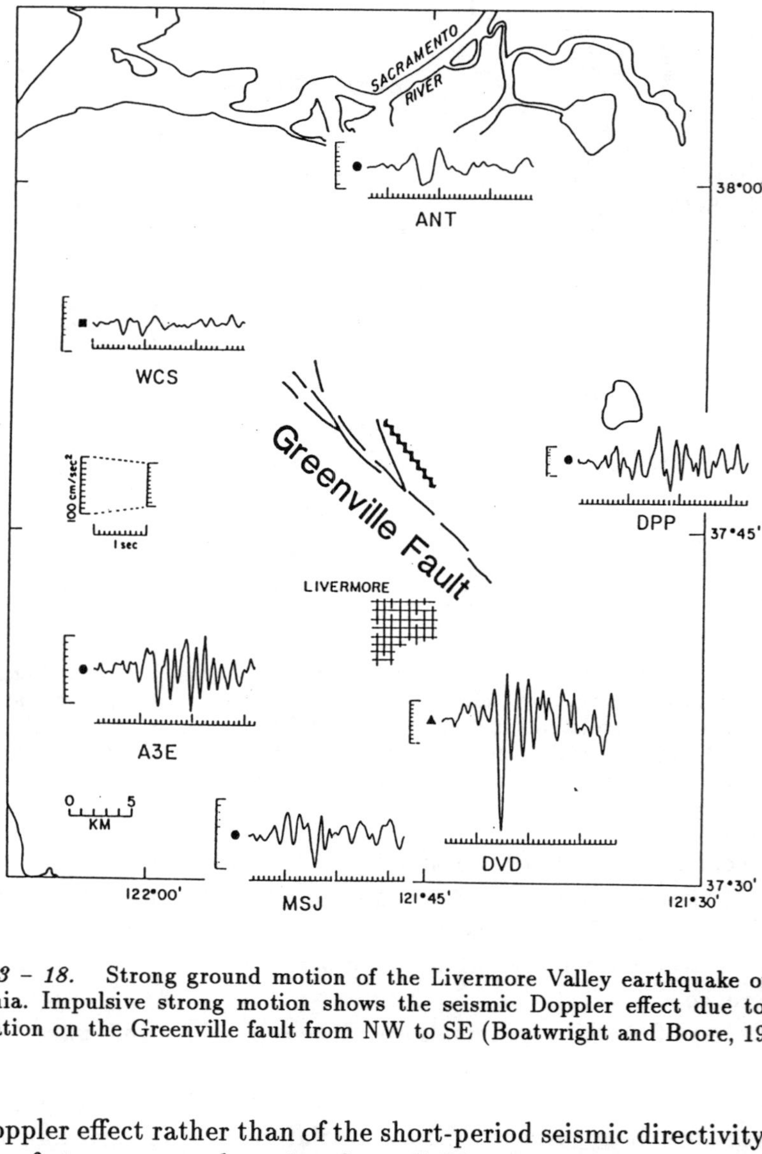

Figure 3 – 18. Strong ground motion of the Livermore Valley earthquake of 1980 in California. Impulsive strong motion shows the seismic Doppler effect due to rupture propagation on the Greenville fault from NW to SE (Boatwright and Boore, 1982).

mic Doppler effect rather than of the short-period seismic directivity. Wavelengths of strong ground motion from Californian earthquakes are comparable to the fault lengths, because the earthquakes analyzed were not very large (magnitude of 6 to 6.5). Examples of strong ground motion for a Californian earthquake are shown in Fig. 3-18. Impulsive strong motion is evidence for the seismic Doppler effect rather than the short-period seismic directivity effect. We notice large difference between the impulsive strong motions in Fig. 3-18 and the complicated strong motion accelerations in Fig.

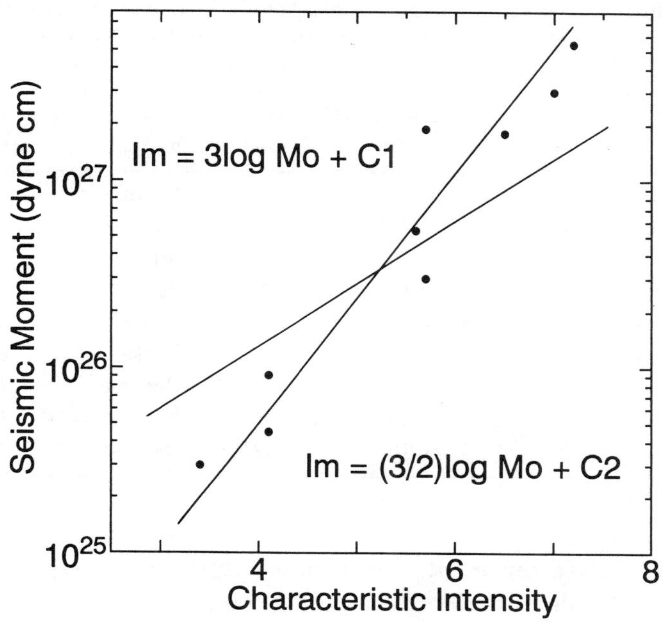

Figure 3 – 19. **Characteristic intensity and seismic moment of disastrous earthquakes in Japan and in China.**

3-2. Strong ground motion is not always due to short waves compared with the earthquake source size. Short-period seismic directivity is valid only for random-phase short waves, so wavelengths must be much shorter than the size of the earthquake source for valid application of this technique.

We have seen that seismic intensity distribution gives information on the kinematic faulting mode of disastrous earthquakes. The faulting mode, fault trend, and rupture velocity can be estimated from the observed distribution of seismic intensities, since the shape of isoseismals is sensitive to these faulting parameters. Since seismic intensities have been investigated in detail for some historical earthquakes, this analysis technique provides potential for retrieving information on the detail source process of such earthquakes.

Geological and local structures do, of course, have effects on seismic intensities, but the local structural effects can be smoothed out by studying isoseismals of smaller intensities over wider areas. Rupture velocities estimated from isoseismals are usually between 70% and 90% of S-wave velocity. All the estimated rupture velocities are larger than those obtained from long-period surface and body waves. Both the short-period and long-period estimates of rupture velocities have uncertainties. Qualitatively, the

TABLE 3 – 1. Disastrous earthquakes studied for characteristic intensity and excitation of short-period accelerations. Faulting mode U (unilateral) or B (bilateral) is estimated from seismic intensity distributions. Characteristic intensity is defined as modified Mercalli intensity at 100 km perpendicular to the fault strike.

Event	M_s	Seismic moment (dyne·cm)	Faulting mode	Slip motion	Characteristic intensity
1983 Japan Sea	7.8	5.4×10^{27}	U	Reverse	7.2
1964 Niigata	7.5	3.0×10^{27}	B	Reverse	7.0
1976 Tangshan	7.7	1.8×10^{27}	B	Strike	6.5
1973 Luhuo	7.5	1.9×10^{27}	U	Strike	5.7
1970 Tonghai	7.5	5.4×10^{26}	B	Strike	5.6
1975 Haicheng	7.4	3.0×10^{26}	U	Strike	5.7
1976 Songpan	6.9	9.1×10^{25}	U	Reverse	4.1
1967 Luhuo	6.1	4.5×10^{25}	U	Normal	4.1
1976 Songpan	6.4	3.0×10^{25}	U/B	Reverse	3.4

long-period estimate represents the average rupture process of the entire faulting, while the short-period estimate represents jerky onsets and abrupt terminations of small-scale fault patches. Thus the difference between the rupture velocities from long-period and short-period seismic waves may be due to the difference in the macroscopic and stochastic faulting processes.

Quantitative representation of seismic intensity is made by measuring modified Mercalli intensity at 100 km perpendicular to the fault strike. This intensity value is designated the characteristic intensity of the earthquake. Since the JMA intensity scale is used for the 1983 Japan Sea earthquake, the observed value of $I_J = 5.0$ is converted to $I_M = 7.2$ using (3-19) and (3-22). Characteristic intensity thus defined has been evaluated for the disastrous earthquakes listed in Table 3-1. Figure 3-19 shows the relation between seismic moment and characteristic intensity of those earthquakes. Since the maximum acceleration is due to the high-frequency source spectrum of the complex faulting process, we can establish the relation in Fig. 3-19.

For the maximum acceleration proportional to seismic moment, the model seismic intensity I_m, from (3-19), is

$$I_m = 3 \log M_o + C_1, \tag{3 - 23}$$

where C_1 is a constant. This is the long-period approximation described in Appendix E. For the maximum acceleration proportional to the square root of seismic moment, it is

$$I_m = \frac{3}{2} \log M_o + C_2, \tag{3 - 24}$$

where C_2 is a constant. This is the short-period approximation of the source spectrum of the complex faulting process studied in §2. The latter case leads to the stochastic similarity. Although the data points are scattered in Fig. 3-19, the last relation (3-24) with $C_2 = -24.0$, is more consistent with the observations than is (3-23). Consequently, the excitation of short-period acceleration can be well represented by the high frequency source spectrum of the complex faulting process together with the stochastic similarity.

CHAPTER 4

SEISMIC ENERGY OF COMPLEX FAULTING PROCESS

Energy is the most fundamental quantity for measuring the strength of earthquake sources. There are basically three different methods used to estimate the energy of earthquakes. The first method gives an empirical estimate for the seismic energy E_S from surface-wave magnitude M_S * by using Gutenberg and Richter's relation:

$$\log E_S = 1.5M_S + 11.8, \qquad (4-1)$$

where E_S is measured in erg. Although this relation has been used for a long time, the physical significance of this method is somewhat uncertain. The second method is a static estimate of the seismic energy obtained from seismic moment M_o (stress drop $\Delta\sigma_0$) and fault dimension LW as the work done by faulting. This gives the energy available for seismic radiation from the faulting process. The third method is a kinetic estimate of the energy in seismic waves.

Theoretically, the kinetic estimate of seismic-wave energy requires knowledge of the space-time history of the faulting process, whereas the static estimate depends only on static parameters of the fault. These two representations are independent of each other and together specify the dynamical boundary condition for the faulting process.

The estimate of seismic energy of an earthquake is usually made under an implicit assumption of smooth-and-coherent faulting. Recent advances in seismometry enable us to calculated seismic-wave energy of large earthquakes from broad-band seismic waves more precisely than before. The results can not be directly applied to understand the complex faulting process of heterogeneous earthquakes, since there is no fundamental theory which links the observation and the heterogeneous earthquake source. In the following, we will derive theoretical expressions for static energy and kinetic seismic-energy generated by the complex faulting process.

4.1. STATIC ESTIMATE OF SEISMIC ENERGY OF HETEROGENEOUS FAULTING

We have stress drop $\Delta\sigma(\vec{\xi})$ and associated dislocation $D(\vec{\xi})$ at a point $\vec{\xi}$ on a fault area S. Over the fault plane, the average stress drop is $\Delta\sigma_0$ and the average dislocation is D_0. Seismic moment M_o in this case is expressed

$$M_o = \mu \int_S D(\vec{\xi})dS = \mu D_0 S. \qquad (4-2)$$

*M_S is the surface-wave magnitude defined by Gutenberg (1945). Recent surface-wave magnitude M_s is slightly different from M_S, because of historical changes in seismometry. We neglect this difference in the following analysis, because the difference is only about 0.2 (Abe, 1981).

Consider a simple model of an earthquake source where $\sigma_1(\vec{\xi})$, $\sigma_2(\vec{\xi})$, and $\sigma_f(\vec{\xi})$ are the initial, final, and frictional stresses at a point $\vec{\xi}$ on the fault. Then the energy available for seismic waves is

$$W_0 = \int_S \left\{ \frac{\sigma_1(\vec{\xi}) + \sigma_2(\vec{\xi})}{2} - \sigma_f(\vec{\xi}) \right\} D(\vec{\xi}) dS. \qquad (4-3)$$

The first term in the integrand indicates the total work done by the stress drop on the fault, and the second is the energy dissipated by the friction on the fault. Since the local stress drop is defined

$$\Delta\sigma(\vec{\xi}) = \sigma_1(\vec{\xi}) - \sigma_2(\vec{\xi}), \qquad (4-4)$$

we can rewrite (4-3) as

$$W_0 = \int_S \left\{ \frac{\Delta\sigma(\vec{\xi})}{2} - (\sigma_f(\vec{\xi}) - \sigma_2(\vec{\xi})) \right\} D(\vec{\xi}) dS. \qquad (4-5)$$

For simplicity we assume Orowan's condition, namely that the fault motion ceases when the stress decreases to the frictional stress. Thus

$$\sigma_2 = \sigma_f. \qquad (4-6)$$

Let $D(\vec{\xi}) = D_0$ and $\Delta\sigma(\vec{\xi}) = \Delta\sigma_0$ everywhere over the fault plane, the energy available for seismic-wave radiation in (4-5) is approximated as

$$\bar{W}_0 = \frac{\Delta\sigma_0 M_o}{2\mu}. \qquad (4-7)$$

This is the second method to estimate the seismic energy of an earthquake. This relationship has been used to estimate the energy released by great shallow earthquakes. For typical observations of $\Delta\sigma_0 = 30$ bar and $\mu = 4 \times 10^{11}$ dyne/cm^2, the above relation yields

$$\bar{W}_0 \simeq (4 \times 10^{-5}) M_o. \qquad (4-8)$$

Applying an empirical relation between seismic moment and surface-wave magnitude, $\log M_o = 1.5M_S + 16.1$, we have

$$\log \bar{W}_0 = 1.5M_S + 11.7, \qquad (4-9)$$

which is consistent with Gutenberg and Richter's energy estimate in (4-1).

The total sum of dislocations on the fault is expressed as the seismic moment in (4-2). Irrespective of how the dislocation fluctuates on the fault plane, the average dislocation D_0 does not change. Therefore, the seismic moment of an earthquake is not affected by random fluctuations of dislocations on the fault. This understanding may be explained another way: The seismic moment relates to the statistical first moment of dislocations summed up on the fault, while the random fluctuation of dislocations is evaluated as the second moment, the variance. Statistically, the first moment is independent of the second moment. We will see that knowledge of the variance of dislocations on the fault is essential for determining the energy of heterogeneous faulting.

When the stress drop on the fault is no longer constant, the theory is not as simple as (4-7). Since the stress drop and dislocation are fluctuating locally on the fault patches, they are approximated by the sum of the deterministic part of the average faulting and the stochastic part of random fault patches as

$$\begin{aligned} \Delta\sigma(\vec{\xi}) &= \Delta\sigma_0 + \delta\sigma(\vec{\xi}), \\ D(\vec{\xi}) &= D_0 + \delta D(\vec{\xi}), \end{aligned} \qquad (4-10)$$

where $\delta\sigma$ and δD represent random fluctuations of local stress drop and local dislocation, respectively, and they are characterized by random variables with zero mean. In this case, the energy W_0 in (4-5) is rewritten as

$$W_h = \bar{W_0} + \frac{1}{2}\iint_S \delta\sigma(\vec{\xi})\delta D(\vec{\xi})dS, \qquad (4-11)$$

where Orowan's condition in (4-6) is assumed. The second term in the right hand side of (4-11) results from the work done by random fault patches. Since the dislocation $D(\vec{\xi})$ on a particular fault patch is large where the net stress drop $\Delta\sigma(\vec{\xi})$ is large, and conversely, the product of $\delta\sigma$ and δD integrated over the fault plane does not vanish.

The Cauchy-Schwarz inequality yields an estimate of the work done by the stochastic part of the complex faulting process. The theory is developed in Appendix D. The upper bound of the energy available for seismic-waves from the complex faulting process is, from (4-7) and (D-7),

$$W_h = \frac{\Delta\sigma_0}{2\mu}M_o\Big\{1 + \frac{<\Delta\sigma^2>}{{\Delta\sigma_0}^2}\Big(\frac{\lambda T_0}{2}\Big)^{1/2}\Big\}. \qquad (4-12)$$

Although the ratio of variance stress drop to the square of average stress drop is small, the second term in the right hand side of (4-12) becomes larger as rise time T_0 (which depends on fault width) becomes longer. This term is not negligible for large earthquakes. The representation of static energy estimate in (4-12) for the complex faulting process reduces to that for a smooth and coherent fault in (4-7) when the fluctuation of local stress drops becomes zero. It will be shown in §6 that the stochastic part of the complex faulting process is significant in determining the seismic energy of natural earthquakes and that the expression of (4-7) for smooth and coherent faulting is not appropriate for the energy of large earthquakes.

4.2. KINETIC ESTIMATE OF SEISMIC-WAVE ENERGY OF HETEROGENEOUS FAULTING

Seismic-wave energy can be evaluated from observed body-waves. Advances in broad-band seismometry and in computational facility enable routine estimations of the seismic-wave energy radiated by natural earthquakes. Table 1-1 in §1 shows some results of these analyses. These estimates of seismic energy have been studied in terms of a smooth-coherent rupture for earthquake sources, but the sources of large earthquakes are neither smooth nor coherent. In order to understand radiated seismic energy in terms of the heterogeneous earthquake source, we must derive a theoretical expression for seismic-wave energy radiated from the complex faulting process.

Seismic-wave energy radiated from the faulting process can be generally expressed:

$$E_S = \rho \int_{-\infty}^{\infty} \int_0^{2\pi} \int_0^{\pi} \{\alpha \dot{u}_r^2 + \beta(\dot{u}_\theta^2 + \dot{u}_\phi^2)\}\ dt\ {r_0}^2 \sin\theta\ d\theta d\phi, \qquad (4-13)$$

where ρ is density, α and β are P- and S-wave velocities. Displacement components u_r, u_θ, and u_ϕ in (2-1) and (2-2) are applied to evaluate the theoretical P- and S-wave energies of the complex faulting process, where the dot stands for the time derivative. The time-domain integral in (4-13) can be converted to the frequency domain by Parseval's equality;

$$\int_{\infty}^{\infty} \dot{u}^2(t)dt = \frac{1}{2\pi} \int_{-\infty}^{\infty} |\hat{\dot{u}}(\omega)|^2 d\omega,$$

where $\hat{\dot{u}}$ is the Fourier transform of $\dot{u}$. This indicates that the total energy in the time domain is equal to that evaluated in the frequency domain. We need Parseval's equality because the source spectrum of the complex faulting process, represented by (2-8) to (2-10), is known. A lengthy but straightforward integration yields

$$\frac{1}{2\pi} \int_{-\infty}^{\infty} |\hat{\dot{u}}_c(\omega)|^2 d\omega \simeq \frac{1}{2\pi} \Big(\frac{M_o |R_{\theta\phi}^c|}{4\pi\rho r_0 c^3}\Big)^2 \Big[\frac{{\omega_c}^2}{2T_0} + \frac{\lambda < \Delta\sigma^2 >}{4\Delta{\sigma_0}^2} {\omega_c}^2\Big], (4-14)$$

where $\lambda > \omega_c$ is assumed, λ is patch corner frequency defined in (2-6) and ω_c is corner frequency defined in (2-11).

The integration of (4-13) by θ and ϕ is now straightforward, yielding P-wave energy radiated from the complex faulting process as

$$E_S^P = \rho\alpha \frac{2\bar{v}^2 {M_o}^2}{\pi\rho^2\alpha^6 L^2}\Big[\frac{1}{2T_0} + \frac{\lambda < \Delta\sigma^2 >}{4\Delta\sigma_0^2}\Big] \times \Big[\frac{2}{3\nu^2} - \frac{4}{\nu^4} + (\frac{1}{\nu^3} - \frac{2}{\nu^5})\ln\Big|\frac{1-\nu}{1+\nu}\Big|\Big], \qquad (4-15)$$

where $\nu = \dfrac{\bar{v}}{\alpha}$, and S-wave energy as

$$E_S^S = \rho\beta \frac{\bar{v}^2 {M_o}^2}{2\pi\rho^2\beta^6 L^2}\Big[\frac{1}{2T_0} + \frac{\lambda < \Delta\sigma^2 >}{4\Delta\sigma_0^2}\Big] \times \Big[\frac{8\chi^4 - 50\chi^2 + 48}{3\chi^4(1-\chi^2)} + \frac{8-3\chi^2}{\chi^5}\ln\Big|\frac{1-\chi}{1+\chi}\Big|\Big], \qquad (4-16)$$

where $\chi = \dfrac{\bar{v}}{\beta}$. The total seismic-wave energy E_S is then

$$E_S = E_S^P + E_S^S. \qquad (4-17)$$

As for the static estimate of seismic energy, the variance stress drop contributes significantly to the seismic energy of the complex faulting process. This is more evident for larger earthquakes for which rise time T_0 becomes longer. Note that the contribution of the stochastic part of the complex faulting process to the static estimate of seismic energy of (4-12) is proportional to $(\lambda T_0)^{1/2}$, whereas for the kinematic estimate of (4-17) it is proportional to λT_0. The seismic-wave energy in (4-17) is identical to the result by Haskell (1964) for smooth-coherent faulting without heterogeneities, that is for $< \Delta\sigma^2 >= 0$.

These theoretical representations are the fundamental for relating the complex faulting process of large earthquakes to the calculated seismic energies. Before investigating the energy budget of natural earthquakes on the basis of the complex faulting process, we consider seismic wave energy and the faulting mode.

The energy ratio of S- to P-waves is strongly dependent on the earthquake faulting process. For example, a double-couple point source gives an energy ratio of about 23.4 and the circular faulting model in §1.6.2 by Sato and Hirasawa (1973) gives about 20. A coherent rupture propagating on a rectangular fault yields a ratio ranging about 30 to 50 with increasing rupture velocity.

When we consider the energy ratio of P- and S-waves of the complex faulting process, the representations in (4-15) and (4-16) predict energy ratios of about 35, 47, 54 and 60, when $\frac{<\Delta\sigma^2>}{\Delta\sigma_0^2}$ is assumed to be 0, $\frac{1}{\lambda T_0}$, $\frac{2}{\lambda T_0}$ and $\frac{4}{\lambda T_0}$, respectively. Corresponding values of $\frac{\sigma^2}{\bar{a}^2}$ to $\frac{<\Delta\sigma^2>}{\Delta\sigma_0^2}$ in the above are 0, 0.5, 1.0 and 2.0, respectively. For this calculation, $L = W$, $\frac{\bar{v}}{\beta} =$ 0.83 and $\frac{\lambda}{(W/\beta)} = 200$ are assumed. The first case of $\frac{<\Delta\sigma^2>}{\Delta\sigma_0^2} = 0$ is for the deterministic source without fault heterogeneities. Increasing variance stress drop and variance dislocation velocity yield increasing energy ratios: the complex faulting process radiates much more S-wave energy than do the other source models.

Although the energy ratios above are based on model parameters rather than observations, they illustrate the effective excitation of short-period S-waves due to the rupture process of small-scale random fault patches. Generally, strong ground motion for large earthquakes is dominated by short-period S-waves as in Fig. 3-2. Rupture process of random fault patches provide one mechanism for that.

4.3. SEISMIC EFFICIENCY OF HETEROGENEOUS FAULTING

4.3.1. *Previous Estimates of Seismic Efficiency*

Seismic efficiency is defined as the ratio of seismic-wave energy (E_S) to total work done (W_t) by an earthquake. That is

$$\eta_0 = \frac{E_S}{W_t}. \tag{4 - 18}$$

The total work is not the same as the energy available for seismic waves in (4-3). It is defined

$$W_t = \int_S \frac{\sigma_1(\vec{\xi}) + \sigma_2(\vec{\xi})}{2} D(\vec{\xi})\, dS, \tag{4 - 19}$$

where S is the earthquake source size. If earthquake faulting is smooth and coherent, local stresses on the fault, $\sigma_1(\xi)$ and $\sigma_2(\xi)$ can be approximated by the corresponding average values of $\bar{\sigma}_1$ and $\bar{\sigma}_2$. Then from (4-5) and (4-19), the coherent seismic efficiency is

$$\bar{\eta}_0 = \frac{(\bar{\sigma}_1 - \bar{\sigma}_2) - 2(\bar{\sigma}_f - \bar{\sigma}_2)}{(\bar{\sigma}_1 + \bar{\sigma}_2)}, \tag{4 - 20}$$

where $\bar{\sigma}_f$ is average frictional stress on the fault. Note that the seismic efficiency of (4-20) does not depend on the earthquake source size. Therefore, any variation of $\bar{\eta}_0$ with respect to the earthquake source size must result from variation in the dynamical conditions of faulting. Applying Orowan's condition again, the efficiency is

$$\bar{\eta}_1 = \frac{\bar{\sigma}_1 - \bar{\sigma}_2}{\bar{\sigma}_1 + \bar{\sigma}_2}, \qquad (4-21)$$

The coherent seismic efficiency in (4-21) is 1, for the complete stress drop ($\bar{\sigma}_2$ =0). For frictional stress greater than 0, that is $\bar{\sigma}_2 \neq 0$, the seismic efficiency is reduced: if stress drop $\bar{\sigma}_1 - \bar{\sigma}_2 = 30$ bar and applied stress $\bar{\sigma}_1 + \bar{\sigma}_2 = 300$ bar, the coherent seismic efficiency is 0.1.

Consider the surface energy necessary for creating new surface of an extending fault plane. Seismic efficiency (4-21) is then rewritten

$$\bar{\eta}_2 = \frac{\bar{\sigma}_1 - \bar{\sigma}_2}{\bar{\sigma}_1 + \bar{\sigma}_2}\Big\{1 - \frac{2\gamma}{(\bar{\sigma}_1 - \bar{\sigma}_2)D_0}\Big\}, \qquad (4-22)$$

where γ is surface energy at the rupture front necessary to create new surface. The surface energy may be estimated from the product of van der Waal's force and atomic distance, which are about 10^{11}dyne·cm^{-2} and about 10^{-8}cm. Then, γ is estimated to be of the order 10^3erg·cm^{-2}. Compared with the average stress drop of 30 bar (3×10^7 dyne·cm^{-2}), the energy required to create new surface is negligibly small.

4.3.2. *Seismic Efficiency of Heterogeneous Faulting*

For irregular earthquake faulting, stress drop and dislocation on the fault change place to place. The upper bound of the total work (4-19) for the complex faulting process can be evaluated as

$$W_t \simeq \frac{\Delta\sigma_0}{2\mu}M_o\Big\{\frac{\bar{\sigma}_1 + \bar{\sigma}_2}{\Delta\sigma_0} + \frac{<\Delta\sigma^2>}{{\Delta\sigma_0}^2}\Big(\frac{\lambda T_0}{2}\Big)^{1/2}\Big\}. \qquad (4-23)$$

This can be derived similarly as (4-12) using (4-3) and (4-11).

Since S-wave energy is dominant, seismic-wave energy radiated from the complex faulting process can be approximated by the S-wave energy of (4-16). Then the seismic efficiency of the complex faulting process is expressed approximately as

$$\eta_h = \frac{p}{4q}\Big\{1 + \frac{<\Delta\sigma^2>}{{\Delta\sigma_0}^2}\frac{\lambda T_0}{2}\Big\}\chi^3 g(\chi)\Big/\Big\{\frac{\bar{\sigma}_1 + \bar{\sigma}_2}{\Delta\sigma_0} + \frac{<\Delta\sigma^2>}{{\Delta\sigma_0}^2}\Big(\frac{\lambda T_0}{2}\Big)^{1/2}\Big\}, \qquad (4-24)$$

where $g(\chi)$ is the function of the ratio of average rupture velocity and S-wave velocity described in (3-3). This representation describes the seismic efficiency in terms of kinematical faulting and heterogeneous faulting parameters.

When earthquake faulting is smooth and coherent, that is, $< \Delta\sigma^2 >= 0$, the above seismic efficiency reduces to

$$\bar{\eta}_r = \frac{\bar{\sigma}_1 - \bar{\sigma}_2}{\bar{\sigma}_1 + \bar{\sigma}_2} \frac{p}{4q} \chi^3 g(\chi). \tag{4 - 25}$$

The above representation should be compared to the seismic efficiency evaluated for the circular crack model (§1.6.2) by Sato and Hirasawa(1973)

$$\bar{\eta}_c = \frac{\bar{\sigma}_1 - \bar{\sigma}_2}{\bar{\sigma}_1 + \bar{\sigma}_2} \frac{3}{7\pi} h(\frac{\alpha}{\bar{v}}, \frac{\beta}{\bar{v}}), \tag{4 - 26}$$

where h is a function of ratios of P- and S-wave velocities to rupture velocity. These are different from the coherent seismic efficiency in (4-20), since the kinematical rupture propagation has been considered in the latter cases of (4-25) and (4-26).

Since the geometrical similarity and kinematical similarity suggest p and q in (4-25) are constant, both cases above yield a seismic efficiency which depends on rupture velocity and is independent of the size of the earthquake.

In contrast, the effect of the earthquake source size appears in the seismic-efficiency representation (4-24) of the complex faulting process in T_0 for the seismic wave energy and in ${T_0}^{1/2}$ for the total work. Since T_0 is related to the size of earthquake sources, this indicates that the seismic efficiency of the complex faulting process increases with the source size. The important effect of fault heterogeneities on seismic energy and seismic efficiency will be fully discussed in §6.

CHAPTER 5

EARTHQUAKE MAGNITUDE AND COMPLEX FAULTING PROCESS

Earthquake magnitude has been determined for earthquakes since the end of the 19th century. More recently magnitude of local earthquakes has also been determined, using a variety of methods with different kinds of data. The current trend in seismology is to describe earthquakes in terms of the source parameters discussed previously. However, earthquake magnitude is a parameter widely used as a measure of both large and small earthquakes and is the only parameter for comparing the strength of historical earthquakes with recent ones. We will study earthquake magnitude, considering the long-period and the short-period approximations of the complex faulting process. The long-period approximation describes the theoretical relationship between earthquake magnitude and macroscopic source parameters, whereas the short-period approximation clarifies the relationship between magnitude and stochastic source parameters.

The seismic moment is an important parameter for estimating the total work done by the faulting; the fault dimension determines the area size over which the strain energy is released; and the earthquake magnitude is a measure of the seismic-wave energy in a particular frequency band. Relations among these three quantities are especially important for understanding the energy budget of the earthquake source process.

The treatment here does not follow the same approach to earthquake magnitude as in previous studies, in which an implicit assumption has been employed, namely that the time-domain amplitude of seismic waves is proportional to the earthquake source spectrum. For example, Kanamori and Anderson (1975) assumed a relation for earthquake magnitude in terms of earthquake source parameters as

$$M_S \propto \log|\Omega(\omega)|,$$

where $|\Omega(\omega)|$ is the earthquake source spectrum similar to (1-40).

The earthquake magnitude is always determined by using the maximum amplitude of seismic waves on seismograms (§1.7), that is earthquake magnitude is measured in the time domain. Consequently, it is necessary to consider the relationship between the time-domain amplitude and the spectral amplitude before applying the above relation to the observation. The maximum amplitude is not a quantity which can be described by standard statistical theory. It is an extreme value in the statistical sense. In the above relation, the physical significance of the maximum amplitude in the time domain has not been considered. Omission of these considerations by the previous studies is fatal for understanding the physical significance of the earthquake magnitude in a general manner.

We will now study earthquake magnitude in more detail, based on its original definition, by considering the band-limited nature of seismic waves and the maximum amplitude in the time domain. Relationships between time-domain amplitude and frequency spectrum will be also developed for continuous-phase long waves and for random-phase short waves.

5.1. LONG-WAVELENGTH DESCRIPTION OF SURFACE-WAVE MAGNITUDE

The definition of surface-wave magnitude states that M_S is the logarithm of the maximum amplitude of dominant surface waves, with the period about 20 sec, after correcting for the wave-travel distance. The wavelength of 20-sec surface waves is about 60 to 80 km. Thus M_S is the measure of a long-wavelength description for those earthquakes with characteristic fault size smaller than 60 km, and conversely is a short-wavelength description for great earthquakes with characteristic fault size much larger than 80 km. An earthquake of 60 to 80 km fault-length has M_S of about 7.5, while $M_S = 8$ corresponds to fault length of more than 200 km. Thus surface-wave magnitude larger than 7.5 represents the short-wavelength measure of earthquake sources and the magnitude smaller than 7.5 describes the long-wavelength measure.

From the source spectrum in (2-8), the approximation for the source spectral density in the extremely low-frequency range yields $LWT_0\bar{a}$. This quantity is equivalent to $\frac{M_o}{\mu}$. Usually the spectrum in this low-frequency range is obtained from long waves with continuous phase. Waves with continuous phase are, of course, amplitude-additive. The time-domain amplitude in a particular frequency band multiplied by the time duration of its wave train is related to the corresponding spectral component. This derivation is given in Appendix E. It is assumed here that the time T_d in (3-12) of $\frac{L}{\bar{v}}$ represents the time duration of the wave train. If the time duration is very short compared with 20 sec, the time-domain amplitude A_{max} of continuous-phase long waves, after correcting for the attenuation due to wave travel-distance and anelasticity, is expressed as (E-5) in Appendix E

$$A_{max} = \frac{c_0 M_o}{2\pi\mu}, \qquad (5-1)$$

where c_0 is a constant independent of the earthquake source size. This is valid for small earthquakes, since the time duration $\frac{L}{\bar{v}}$ is assumed to be short compared with 20 sec.

For large earthquakes, where $\frac{L}{\bar{v}}$ is not short compared with 20 sec, the amplitude of continuous-phase long waves is approximated by the seismic

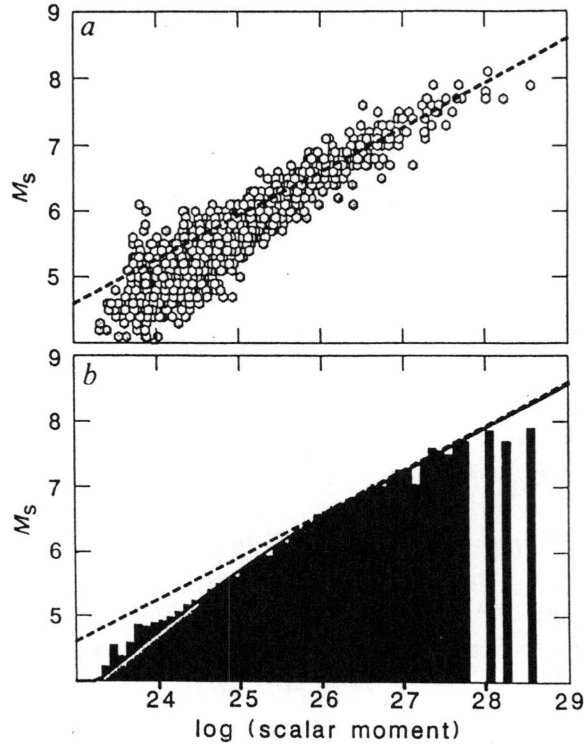

Figure 5 − 1. Surface-wave magnitude M_s and seismic moment M_o (in dyne·cm) for recent major earthquakes (Ekström and Dziewonski, 1988). (a) shows individual data points, and (b) indicates observations averaged over 0.1-unit ranges of log M_o. The dashed line indicates the moment-magnitude relation for large earthquakes. Solid lines are for small moment range and for large moment range.

moment divided by the time duration as (E-7) in Appendix E

$$A_{max} = \frac{c_0 M_o \bar{v}}{\mu L}. \qquad (5-2)$$

We should notice the same constant c_0 in (5-1) and (5-2), although the equations seem to give different dimension.

Since A_{max} in (5-1) and (5-2) is the maximum amplitude in the time domain, the logarithm of A_{max} should be a measure of a theoretical surface-wave magnitude M_S for small earthquakes from (5-1) as

$$M_S = \log M_o - \log(2\pi\mu) + \epsilon, \qquad (5-3)$$

and for large earthquakes from (5-2)

$$M_S = \log\left(\frac{M_o}{L}\right) + \log\left(\frac{\bar{v}}{\mu}\right) + \epsilon. \qquad (5-4)$$

Here ϵ is a constant depending on the magnitude definition and the unknown constant c_0. The assumption of $T_d \simeq \frac{L}{\bar{v}}$ employed for the time duration is heuristic and must be justified from observations, however, the term ϵ can compensate for error in this assumption.

The dimension of spectrum is that of time-domain amplitude multiplied by time. Most previous studies assumed the source spectrum as a measure of surface-wave magnitude as was shown in the opening paragraph of this chapter. The differences in (5-3) and (5-4) suggests that the assumption of the previous studies is not appropriate. The theoretical representations of surface-wave magnitude in (5-3) and (5-4) more rigorously follow the original definition of the magnitude scale. This treatment will be extended to cover surface-wave magnitude from random-phase short waves in the next section.

A relationship between seismic moment and surface-wave magnitude for large earthquakes can be derived from (5-4) and (2-15), applying the similarity laws discussed earlier;

$$\log M_o = \frac{3}{2}M_S - \frac{1}{2}\log\left(\frac{p^2\Delta\sigma_0\bar{v}^3}{\mu^3}\right) - \frac{3}{2}\epsilon. \qquad (5-5)$$

Similarly, when the fault area S is given by LW, a theoretical relationship between M_S and S for large earthquakes is

$$\log S = M_S - \log(pq\bar{a}) - \epsilon. \qquad (5-6)$$

The relationship between M_S and M_o in (5-3) for small earthquakes reveals the physical meaning of an empirical relation in Fig. 5-1

$$\log M_o = M_s + 19.24 \qquad M_s < 5.3, \qquad (5-7)$$

where the unit of M_o is dyne·cm. This empirical relation was obtained by recent seismic-moment determinations and by the extended determinations of M_s for small earthquakes facilitated by modern seismological instruments. The empirical relation (5-7) and the corresponding one in (5-3) have the same form. Combining them yields an important constraint on scaling rule of source parameters for small earthquakes:

$$\log(2\pi\mu) - \epsilon = 19.24 \qquad M_s < 5.3, \qquad (5-8)$$

where all parameters are in cgs units due to the unit of M_o. This equation is essential for understanding the physical basis of the earthquake magnitude, a purely empirical parameter.

For large earthquakes we have an empirical relation in Fig. 5-1

$$\log M_o = 1.5M_s + 16.14. \tag{5-9}$$

This empirical relation is the same as that we derived in (5-5) for large earthquakes, and another constraint is obtained

$$\log\left(\frac{p^2 \Delta\sigma_0 \bar{v}^3}{\mu^3}\right) + 3\epsilon = -32.28. \tag{5-10}$$

Thus the empirical relation in (5-9) reflects the dynamical similarity of a constant stress drop $\Delta\sigma_0$, when the geometrical similarity $p = constant$ holds.

Further, an empirical relation between fault dimension and earthquake magnitude, known as Utsu-Seki relation (1954), has been obtained (Fig. 5-2),

$$\log S = 1.02M - 4.01, \tag{5-11}$$

where S (km^2) is from aftershock areas of large earthquakes in and near Japan. This equation was originally derived for the JMA magnitude scale M, however it has been calibrated against M_S. When we round off small quantities, the empirical relation (5-11) is the same as the theoretical relationship derived in (5-6), and the third constraint is

$$\log(pq\bar{a}) + \epsilon = -6.0. \tag{5-12}$$

Thus the empirical relation (5-11) demonstrates the kinematical similarity of $\bar{a}$, when we have the geometrical similarity of p.

These empirical relations between M_o and M_S and between S and M_S, which have been well-known for a long time, are understood here to result from the similarity laws among macroscopic source parameters of the complex faulting process. Note that these understandings are based on the long-wavelength description of the M_S determination.

5.2. SHORT-WAVELENGTH DESCRIPTION OF SURFACE-WAVE MAGNITUDE

The empirical relation in (5-11) between fault dimension S and magnitude M_S fits the data of large earthquakes for M_S smaller than about 7. However, there are systematic deviations of great earthquake data from the relation.

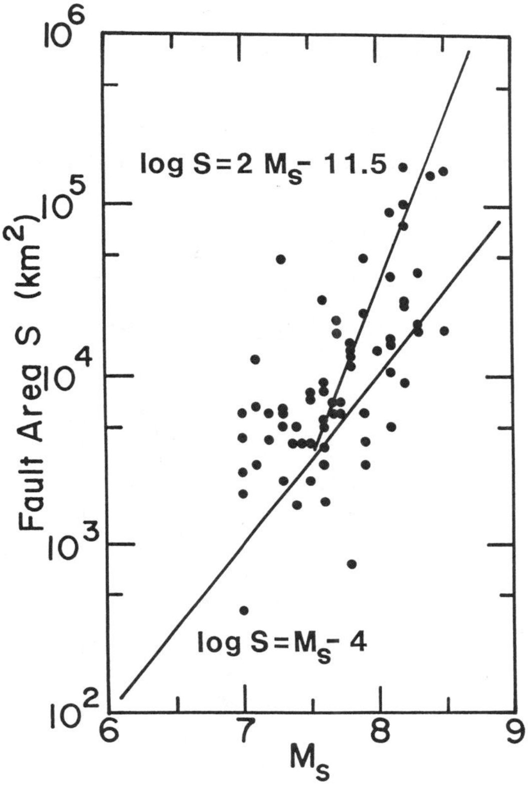

Figure 5 – 2. Relation between surface-wave magnitude M_S and earthquake source dimension S (Koyama and Shimada, 1985). An empirical relation of $\log S = M_S - 4$ is equivalent to Utsu and Seki's (1954) relation for large earthquakes in and near Japan. The other relation, $\log S = 2M_S - 11.5$, is for great earthquakes worldwide.

Figure 5-2 shows M_S and S for great earthquakes worldwide. For these data of great earthquakes we have

$$\log S = 2M_S - 11.5, \qquad (5-13)$$

where S is in km^2. This relation fits reasonably well to the scattered data when M_S is larger than 7.5, for which 20-sec surface-waves are no longer long-wavelength nor long-period measure of the earthquake source.

A similar observation can be seen for the $M_o - M_S$ relation. Figure 5-3 summarizes M_o and M_S for large and great earthquakes. The 1960 Chilean, 1964 Alaskan, 1957 Aleutian, 1952 Kamchatkan, and the 1965 Rat Island earthquakes are the greatest, and they deviate systematically from the previous empirical $M_o - M_S$ relation of (5-9). Another empirical relation is proposed for these data of great earthquakes

$$\log M_o = 3M_S + 4.9. \qquad (5-14)$$

It seems that the data set is too small to confirm this relation definitely, but a similar change in trend was seen in the recent independent data in Fig. 5-1.

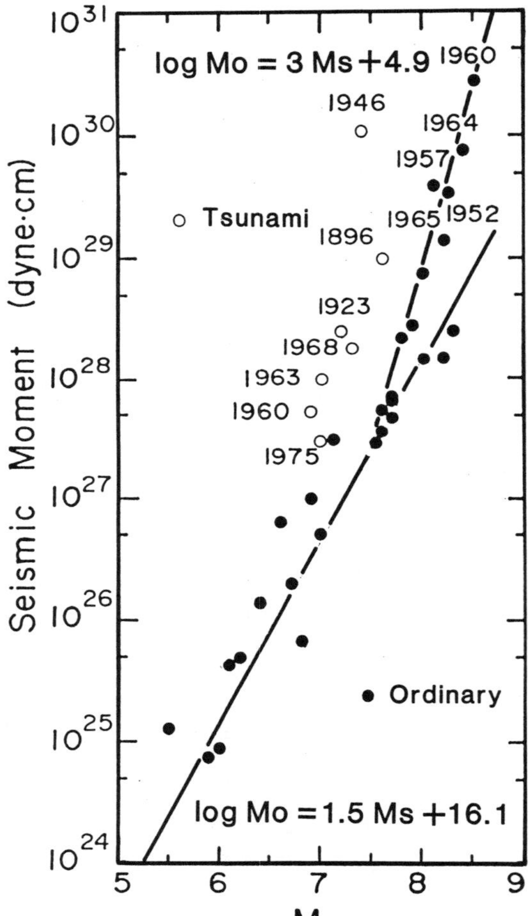

Figure 5 – 3. Seismic moment and surface-wave magnitude for large and great earthquakes (Takemura and Koyama, 1983). Solid and open circles indicates ordinary and tsunami(low-frequency) earthquakes, respectively. Note that large tsunami earthquakes show the similar tendency between M_o and M_S as great earthquakes in the world. Both of them systematically deviate from the general trend of the relation for large earthquakes.

5.2.1. *Destructive Interference Model*

Let us now consider the short-period approximation for the source spectrum, for which there is no heterogeneity ($\sigma^2 = 0$). The particular frequency of $\frac{2\pi}{20}$ for M_S determination is designated ω_s. Consider the case of ω_s much larger than ω_c but smaller than $\frac{2}{T_0}$, where ω_c and $\frac{2}{T_0}$ are the characteristic corner frequencies of the source spectrum in (2-8). This is the case when an approximation of $\left|\frac{\sin(\omega_s/\omega_c)}{(\omega_s/\omega_c)}\right| \simeq \frac{1}{|\omega_s/\omega_c|}$ is valid. This spectral decrease is due to the destructive interference discussed in §2.2. Since the corner frequency of ω_c in (2-11) can be approximated as $\frac{2\bar{v}}{L}$, the source spectrum $|A_c(\omega)|$ at ω_s reduces to $\frac{M_o}{\mu\omega_s L/2\bar{v}}$. In this case, the time domain amplitude A_{max} is similarly to (5-2):

$$A_{max} = c_0 \frac{2\bar{v}M_o}{\mu\omega_s L}\frac{\bar{v}}{L}, \qquad (5-15)$$

where again we assume a time duration $\frac{L}{\bar{v}}$ of wave trains and remind the same dimension of $c = 0$. By taking the logarithm of the maximum amplitude in (5-15) and by applying the similarity laws among macroscopic source parameters, a theoretical surface-wave magnitude can be obtained from (5-15). It would suggest a relationship between seismic moment and surface-wave magnitude for large earthquakes

$$\log M_o = 3M_S - \log\left(\frac{8p^4q^2\bar{a}^2\bar{v}^4}{\mu\omega_s^3}\right) - 3\epsilon \;\left(\frac{20}{\pi} < \frac{L}{\bar{v}} < \frac{20}{\pi pq}\right). \qquad (5-16)$$

This relationship has been used to study the earthquake source parameters of large earthquakes for a long time. The condition attached to (5-16) was derived from the assumption of $\omega_c < \omega_s < \frac{2}{T_0}$, which reduces approximately to $20 < L(km) < 80$, taking $\bar{v} = 3$km/s and $p = q = \frac{1}{2}$. Thus the long-wavelength description of M_S is also appropriate for such earthquakes. This is contrary to our expectations, since M_S is a short-wavelength measure for great earthquakes. Therefore, the previous model by distractive interference of the source spectrum would not explain the observed relation in (5-14).

5.2.2. *Scaling Patch Corner Frequency Model*

A theoretical surface-wave magnitude can be derived from the short-period approximation of the complex faulting process: Consider the source spectrum in (2-8). The rise time T_0 increases in accordance with the size of earthquakes, since it is proportional to $\frac{W}{\bar{v}}$. T_0 for great earthquakes becomes large enough to satisfy a condition:

$$\left(\frac{2}{\omega_s T_0}\right)^2 \ll \frac{2\sigma^2}{{T_0}^2\bar{a}^2}\frac{\lambda T_0}{{\omega_s}^2+\lambda^2}. \qquad (5-17)$$

In this case, 20-sec surface waves are excited primarily by the rupture process of random fault patches rather than by the deterministic part of the complex faulting process. These 20-sec surface waves are characterized as short-waves, and the phase of waves is random. Amplitude-additive spectral analysis is no longer appropriate, rather energy-additive spectral analysis applies. A theory is developed in Appendix F to derive the relationship between time-domain amplitude and spectral amplitude of random-phase short waves. Applying the condition (5-17) and $\omega_c < \omega_s$ to (2-8), the *rms* amplitude of random-phase short waves at ω_s can be expressed (Appendix F)

$$A_{rms} \simeq c_0\left\{\frac{1}{\pi}\left(\frac{2\bar{v}M_o}{\mu\omega_s L}\right)^2\left[\frac{2\sigma^2}{T_0^2\bar{a}^2}\frac{\lambda T_0}{{\omega_s}^2+\lambda^2}\right]\frac{\bar{v}}{L}\right\}^{1/2}, \qquad (5-18)$$

where a time duration of $\frac{L}{\bar{v}}$ is also assumed for random-phase short waves. A constant c_0 is again introduced, which does not depend on the earthquake source size.

Neither a scaling law nor an absolute value of patch corner frequency λ is known for large and great earthquakes. In order to extend the theory of M_S for random-phase short waves, two models for the patch corner frequency are considered. The first model is the same as (2-25) where λ is scaled by the corner frequency even for great earthquakes. This model has been abbreviated as the scaling P-model. In this context, Figure 3-4 suggests ${\omega_s}^2+\lambda^2 \simeq 2{\omega_s}^2$ for large and great earthquakes, since an extrapolation of patch corner frequency gives $f_p(=\frac{\lambda}{2\pi}) \simeq 0.05$Hz for $M_o = 10^{28}$ dyne·cm. By applying the similarity laws among the macroscopic source parameters, the maximum amplitude of random-phase short waves is expressed in terms of the *rms* amplitude in (5-18) as

$$A_{max} \simeq \frac{2c_0\bar{v}}{{\omega_s}^2}\left(\frac{pq\lambda\sigma^2}{\pi}\right)^{1/2} W\ f(N), \qquad (5-19)$$

where $f(N)$ is a function of the number N of peaks and troughs of short waves resulting from the statistics of extreme values. The function $f(N)$ is found in (3-14) and (3-15) corresponding to Rayleigh and Gauss distributions. The parameter N is defined similar to (3-16).

5.2.3. *Constant Patch Corner Frequency Model*

The second model assumes a constant patch corner frequency for great earthquakes. Hereafter this model is designated the constant P-model. Since Figure 3-3 indicates $f_p(=\frac{\lambda}{2\pi})$ as small as 1Hz for small and moderate-size earthquakes, $\lambda = 2\pi$ is tentatively assumed (this will be re-examined more specifically in §6). In this case, the condition ${\omega_s}^2 \ll \lambda^2$ is satisfied, since ω_s is $2\pi/20$. The *rms* amplitude in (5-18) is then rewritten to give the maximum amplitude of random-phase short-waves as

$$A_{max} \simeq \frac{2c_0\bar{v}}{\omega_s}\left(\frac{2pq\sigma^2}{\pi\lambda}\right)^{1/2} Wf(N). \qquad (5-20)$$

The logarithm of A_{max} in (5-19) and (5-20) gives a measure of surface-wave magnitude M_S from random-phase short waves. Consequently, a theoretical surface-wave magnitude of M_S for great earthquakes is

$$M_S = \log W + \log\left(\frac{2\bar{v}}{{\omega_s}^2}\left[\frac{pq\lambda\sigma^2}{\pi}\right]^{1/2} f(N)\right) + \varepsilon \qquad (5-21)$$

for the scaling P-model, and

$$M_S = \log W + \log\left(\frac{2\bar{v}}{\omega_s}\left[\frac{2pq\sigma^2}{\pi\lambda}\right]^{1/2} f(N)\right) + \varepsilon \qquad (5-22)$$

for the constant P-model. This constant ε may be slightly different from ϵ in (5-3), but we assume they are equal because of the approximation employed in this analysis and of the uncertainty involved in the magnitude determination.

Since $\log S = 2\log W - \log p$ when $S = LW$ is assumed, the above theoretical magnitude reduces to

$$\log S = 2M_S - 2\log\left(\frac{2\bar{v}}{{\omega_s}^2}\left[\frac{pq\lambda\sigma^2}{\pi}\right]^{1/2} f(N)\right) - \log p - 2\varepsilon, \qquad (5-23)$$

and

$$\log S = 2M_S - 2\log\left(\frac{2\bar{v}}{\omega_s}\left[\frac{2pq\sigma^2}{\pi\lambda}\right]^{1/2} f(N)\right) - \log p - 2\varepsilon \qquad (5-24)$$

for the scaling and the constant P-models, respectively. These expressions are the short-wavelength description of the complex faulting process for determining the S - M_S relation.

Taking the logarithm of seismic moment $M_o = \Delta\sigma_0 LW^2$, we have

$$\log M_o = 3\log W + \log\left(\frac{\Delta\sigma_0}{p}\right). \qquad (5-25)$$

Substituting the theoretical magnitude in (5-21) and (5-22) into the above equation, a theoretical relationship between M_o and M_S is derived based on the short-period approximation of the complex faulting process:

$$\log M_o = 3M_S - 3\log\Big(\frac{2\bar{v}}{{\omega_s}^2}\Big[\frac{pq\lambda\sigma^2}{\pi}\Big]^{1/2} f(N)\Big) + \log\Big(\frac{\Delta\sigma_0}{p}\Big) - 3\varepsilon \quad (5-26)$$

for the scaling P-model, and for the constant P-model

$$\log M_o = 3M_S - 3\log\Big(\frac{2\bar{v}}{\omega_s}\Big[\frac{2pq\sigma^2}{\pi\lambda}\Big]^{1/2} f(N)\Big) + \log\Big(\frac{\Delta\sigma_0}{p}\Big) - 3\varepsilon. \quad (5-27)$$

Comparing the theoretical representations in (5-23) and (5-24) and the empirical $S - M_S$ relation in (5-13) shows why the empirical relation between fault dimension and surface-wave magnitude for great earthquakes is different from that for large earthquakes. The empirical relation in (5-13) suggests another independent constraint, in c.g.s. units,

$$2\log\Big(\frac{2\bar{v}}{{\omega_s}^2}\ \Big[\frac{pq\lambda\sigma^2}{\pi}\Big]^{1/2} f(N)\Big) + \log p + 2\varepsilon = 1.5, \quad (5-28)$$

$$2\log\Big(\frac{2\bar{v}}{\omega_s}\ \Big[\frac{2pq\sigma^2}{\pi\lambda}\Big]^{1/2} f(N)\Big) + \log p + 2\varepsilon = 1.5, \quad (5-29)$$

for the scaling and the constant P-models, respectively. From the observations this constraint specifies the stochastic similarity $\lambda\sigma^2$.

For the theoretical representations in (5-26) and (5-27) and the empirical $M_o - M_S$ relation in (5-14), we see physically why the empirical relation between seismic moment and surface wave magnitude for great earthquakes is different from those for small and large earthquakes. This understanding provides another constraint:

$$\log\Big(\frac{\Delta\sigma_0}{p}\Big) - 3\log\Big(\frac{2\bar{v}}{{\omega_s}^2}\ \Big[\frac{pq\lambda\sigma^2}{\pi}\Big]^{1/2} f(N)\Big) - 3\varepsilon = 4.9, \quad (5-30)$$

$$\log\Big(\frac{\Delta\sigma_0}{p}\Big) - 3\log\Big(\frac{2\bar{v}}{\omega_s}\ \Big[\frac{2pq\sigma^2}{\pi\lambda}\Big]^{1/2} f(N)\Big) - 3\varepsilon = 4.9, \quad (5-31)$$

for the scaling and constant P-models, respectively. This constraint specifies the dynamical and stochastic similarities.

We have discussed the physical basis of surface-wave magnitude based rigorously on the original definition of the magnitude scale and demonstrated that the wavelength of 20-sec surface waves is very important for understanding the empirical relations between magnitude and source parameters based on the complex faulting process. We derived five independent constraints for the kinematical similarity of $\bar{a}$, the stochastic similarity of $\lambda\sigma^2$, and the constant for the magnitude determination of ϵ. We will determine these parameters in §6 and discuss the general scaling law of the complex faulting process.

5.3. SHORT-PERIOD DESCRIPTION OF BODY-WAVE MAGNITUDE

We now examine body-wave magnitude m_b based on the complex faulting process to provide the support for the theory developed in the preceding section. Body-wave magnitude m_b is calculated from the maximum amplitude of body-waves at a period of about 1 sec* . Therefore, the characteristic frequency ω_b is reasonably assumed to be 2π. In this frequency range, the short-period approximation of the complex faulting process in (5-17) is always satisfied for earthquakes with M_o larger than 10^{20} dyne·cm. For large earthquakes, body waves in this period range are always complicated and are characterized by random phase, as seen in §2.4. The theoretical body-wave magnitude can be derived similarly to (5-22):

$$m_b = \log W + \log\Big(f(N_b)\frac{2\bar{v}}{\omega_b}\Big[\frac{2pq\sigma^2}{\pi\lambda}\Big]^{1/2}\Big) + \varepsilon_b, \qquad (5-32)$$

where $\omega_b = 2\pi$ is used, N_b is the number of peaks and troughs, and ε_b is a constant that depends on the definition of body-wave magnitude and on a propagation constant. This representation is for the constant P-model. The expression for the scaling P-model can also be derived.

A theoretical relationship between seismic moment and body-wave magnitude is obtained by substituting the above equation into (5-25)

$$\log M_o = 3m_b - 3\log\Big(\frac{2\bar{v}}{\omega_b}\Big[\frac{2pq\sigma^2}{\pi\lambda}\Big]^{1/2} f(N_b)\Big) + \log\Big(\frac{\Delta\sigma_0}{p}\Big) - 3\varepsilon_b. (5-33)$$

This theoretical relationship is plotted in Fig. 5-4 using the similarity laws previously discussed. To determine ε_b we use the 1964 Alaskan earthquake, with redetermined m_b of 7.6 and the seismic moment of 7.5 $\times 10^{29}$ dyne·cm. Three relationships are plotted in Fig. 5-4: with function $f(N_b)$ from the statistical theory of extremes in (5-33) assumed to be a constant of 3, and with $f(N_b)$ evaluated similarly (3-14) and (3-15) for Rayleigh and

*Body-wave magnitude m_b is reported routinely by the U.S. Geological Survey and the International Seismological Centre. The magnitude is slightly different from m_B defined and evaluated for large earthquakes by Gutenberg and Richter (1954), because of the historical changes in seismometry.

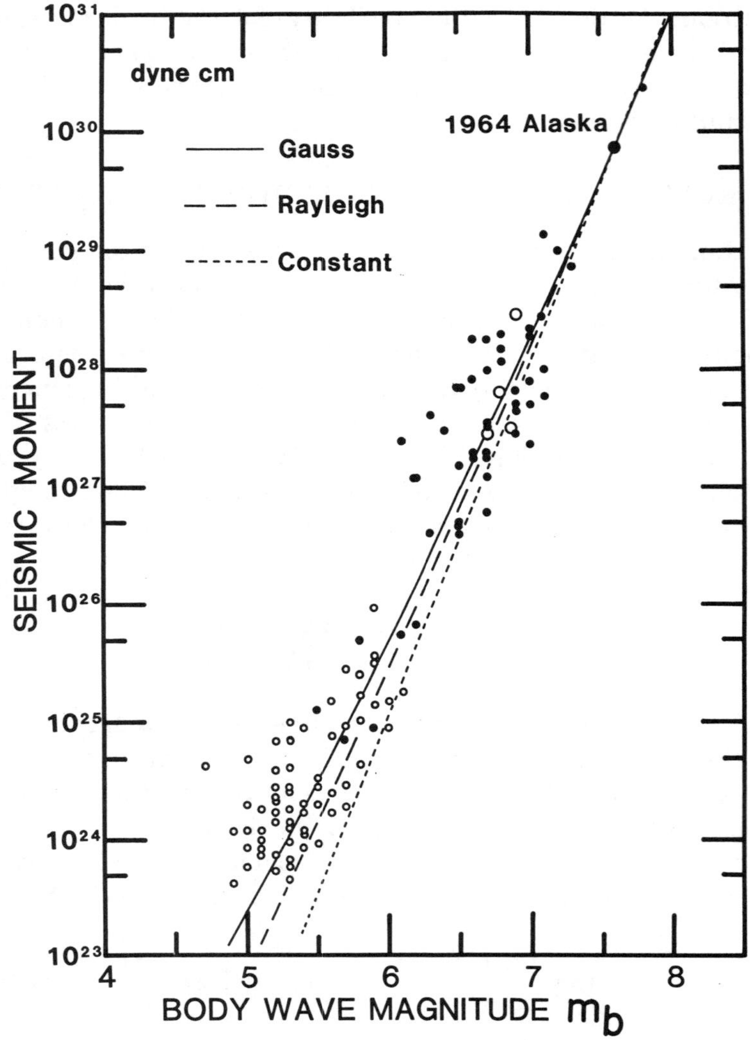

Figure 5 – 4. Relation between body-wave magnitude m_b and seismic moment M_o (Koyama, 1994). The data for small earthquakes (open circles) are from the December 1990 issue of Earthquake Data Report by the U.S. Geological Survey. The data for large and great earthquakes along subduction zones from Houston and Kanamori (1986) are shown by large open circles and from Koyama and Zheng (1985) by solid circles. M_o values are from Lay et al.(1982) and Purcaru and Berckehemer(1982). The Alaskan earthquake with redetermined $m_b = 7.6$ is used as a reference for determining ε_b for the theoretical m_b representation. Theoretical $m_b - M_o$ relationships are evaluated assuming Gauss and Rayleigh distribution for peak and trough amplitudes, and constant of 3 for $f(N)$ irrespective to earthquake sizes.

Gauss distributions respectively as

$$f(N_b) = (2\ln N_b)^{1/2} + \frac{\gamma}{(2\ln N_b)^{1/2}}, \qquad (5-34)$$

and/or

$$f(N_b) = \left\{\ln\left(\frac{N_b{}^2}{2\pi}\right) - \ln\left(\ln\left(\frac{N_b{}^2}{2\pi}\right)\right)\right\}^{1/2}, \qquad (5-35)$$

where, similar to (3-16), the value of N_b in all three cases is

$$N_b = \frac{2\pi L}{\bar{v}}, \qquad (5-36)$$

for a characteristic frequency of $\omega_b = 2\pi$ and a corner frequency of $\omega_c \simeq \dfrac{2\bar{v}}{L}$. Thus the size effect of earthquake source on the body-wave magnitude is represented in terms of $f(N)$ as well as W.

The theoretical $\log M_o - m_b$ relationship in (5-33) for Rayleigh and Gauss distributions explains the observations in Fig. 5-4 over the range from small earthquakes with a seismic moment of 10^{21} to the greatest earthquakes with a seismic moment of 10^{31} dyne·cm. However, the case of $f(N_b) = 3$ does not show the size dependence on earthquake source and systematically deviates from the observations. This is a new finding that the factor of $f(N_b)$ derived from the statistical theory of extremes is very important for the body-wave magnitude.

At a first glance, the data in Fig. 5-4 seems to suggest a relation of $\log M_o = \dfrac{5}{2}m_b + constant$. Many studies have suggested such an empirical relation on the basis of best fit line. However, such an empirical relation does not appear to have any physical significance. Although the coefficient of $\dfrac{5}{2}$ seems to satisfy the observations, it is not a consequence of the earthquake source process. The theoretical body-wave magnitude in (5-33) requires that a coefficient of 3 for the $\log M_o - m_b$ relationship. This relationship is influenced by the properties of random-phase seismic waves which depend on the size of earthquake source. This introduces bias effect resulting in the deceptive coefficient of about $\dfrac{5}{2}$.

Also, similarly to (5-24), we can derive a theoretical relationship between fault dimension and body-wave magnitude. For great earthquakes, that theoretical relationship explains the observed data for $S - m_b^*$, where m_b^* is redetermined as in §2.4.

We have discussed the physical basis for surface-wave magnitude and for body-wave magnitude based on the complex faulting process. The magnitude scale is determined from seismic-wave amplitudes in a particular frequency band, and it is evaluated from the maximum amplitude in the time domain. Therefore the finite wavelength and the statistical theory of extremes are essential in understanding the earthquake magnitude in terms of earthquake source parameters.

CHAPTER 6

GENERAL DESCRIPTION OF COMPLEX FAULTING PROCESS

6.1. SIMILARITY PARAMETERS OF COMPLEX FAULTING PROCESS

The physical basis for the earthquake magnitude has been established in terms of the source parameters of the complex faulting process through the long-period and the short-period approximations. Essential parameters necessary to quantify the complex faulting process in this context are the fault length L, the kinematical similarity of dislocation velocity $\bar{a}$, and the stochastic similarity $\lambda\sigma^2$. The remaining parameters for subduction zone earthquakes can be scaled and estimated from the similarity laws. Therefore, after we determine the kinematical and stochastic similarities, the earthquake source can be evaluated by seismic moment M_o (or fault length L) in the long-period as well as in the short-period ranges.

We have derived constraints for the complex faulting process of natural earthquakes in (5-8), (5-10), (5-12), (5-28) or (5-29), and (5-30) or (5-31). Those constraints are based on well-known empirical relations from abundant worldwide data compiled from classical and modern observations. They are obtained mainly from earthquake magnitudes which are relative scales for measuring the strength of earthquake sources. It is not necessary to allow for unknown whole-path propagation-effects, since the effects are independent of the earthquake source size. We use this advantage in investigating the complex faulting process in a general manner.

The simultaneous equations to be solved are

$$\epsilon = \log(2\pi\mu) - 19.24, \qquad (5-8)$$

$$\log(\bar{a}) + 3\epsilon = -\log\Big(\frac{p^2 q \bar{v}^2}{\mu^2}\Big) - 32.28, \qquad (5-10)$$

$$\log(\bar{a}) + \epsilon = -\log(pq) - 6.0, \qquad (5-12)$$

$$\log(\lambda\sigma^2) + 2\epsilon = -\log\Big(\frac{4\bar{v}^2}{\omega_s^4}\frac{p^2 q}{\pi} f(N)^2\Big) + 1.5, \qquad (5-28)$$

or

$$\log(\frac{\sigma^2}{\lambda}) + 2\epsilon = -\log\Big(\frac{4\bar{v}^2}{\omega_s^2}\frac{p^2 q}{\pi} f(N)^2\Big) + 1.5, \qquad (5-29)$$

and

$$\log(\bar{a}) - \frac{3}{2}\log(\lambda\sigma^2) - 3\epsilon = 3\log\Big(\frac{2\bar{v}}{\omega_s^2}\Big[\frac{pq}{\pi}\Big]^{1/2} f(N)\Big) - \log\Big(\frac{\mu}{\bar{v}}\frac{q}{p}\Big) + 4.9, \quad (5-30)$$

or

$$\log(\bar{a}) - \frac{3}{2}\log(\frac{\sigma^2}{\lambda}) - 3\epsilon = 3\log\Big(\frac{2\bar{v}}{\omega_s}\Big[\frac{pq}{\pi}\Big]^{1/2} f(N)\Big) - \log\Big(\frac{\mu}{\bar{v}}\frac{q}{p}\Big) + 4.9, \quad (5-31)$$

respectively for the scaling P-model and the constant P-model.

We assume rupture velocity $\bar{v} = 3$ km/sec, rigidity $\mu = 4\text{x}10^{11}$ dyne/cm^2, and $p = q = \frac{1}{2}$ for subduction zone earthquakes. Then the similarity parameters of the complex faulting process can be estimated by the least squares method. An average dislocation velocity is thus estimated

$$\bar{a} \simeq 30 \text{ cm/sec}. \tag{6 – 1}$$

The stochastic similarity is evaluated for the scaling P-model as

$$\lambda\sigma^2 \simeq 280 \text{ cm}^2/\text{sec}^3, \tag{6 – 2}$$

and for the constant P-model as

$$\frac{\sigma^2}{\lambda} \simeq 1400 \text{ cm}^2/\text{sec}. \tag{6 – 3}$$

The above two values are evaluated assuming that $f(N) = 2$. The constant ϵ for the definition of surface-wave magnitude is also determined to be

$$\epsilon \simeq -6.86. \tag{6 – 4}$$

Substituting these values from (6-1) to (6-4) in the theoretical representations, we can retrieve exactly the empirical relations among surface wave magnitude M_S, earthquake source area S, and seismic moment M_o previously investigated in §5.

The average dislocation velocity $\bar{a}$ is dependent on the assumption of q, since $q\bar{a}$ is constrained by the above analysis. Although it appears that the stochastic similarity relates to $f(N)$, the values in (6-2) and (6-3) are robust. This is because the number of surface waves in the wave train concerned does not change significantly between large and great earthquakes. The formal representation in (2-14) gives an estimate of stress drop of about 20 bar for the dislocation velocity in (6-1). However, this value is actually about 30 bar, since it is necessary to take into account a geometrical factor of about 1.5 for the stress drop in (2-14), when the ratio of $\frac{L}{W}$ is 2. Therefore, the result in (6-1) also satisfies the constraint on constant stress drop for large subduction-zone earthquakes in Fig. 2-7.

The difficulty in correcting for the attenuation of short-period seismic waves has caused a debate on the use of high-frequency source spectrum for resolving the fault heterogeneities. Body-wave analyses of large earthquakes agree with the Brune's source spectrum for large and great earthquakes. Brune's model (1970) was derived intuitively for a circular fault without heterogeneities. It is obvious that this model is not consistent with long narrow faults of natural earthquakes. The stochastic similarity in (6-2) or (6-3) suggests the importance of small-scale fault heterogeneities on a rectangular fault plane with a geometrical similarity of $\frac{W}{L} \simeq 0.5$. Since the stochastic similarity thus evaluated is obtained from empirical relations for earthquake source parameters on a relative basis, it is free from uncertainties in whole-path propagation effects.

Although the similarity parameters above are typical values for an average earthquake with heterogeneous faulting, nevertheless, it is important to constrain the deterministic part and the stochastic part of the complex earthquake source process. From the fundamental assumption of the present theory in (2-17) and (5-17), we have a criterion for the ratio of variance dislocation velocity σ^2 and average dislocation velocity $\bar{a}$;

$$\frac{4}{\lambda T_0} < \frac{\sigma^2}{\bar{a}^2} < \frac{\lambda T_0}{2}, \tag{6 – 5}$$

and a criterion for variance stress drop;

$$\frac{8}{(\lambda T_0)^2} < \frac{< \Delta\sigma^2 >}{\Delta\sigma_0^2} < 1. \tag{6 – 6}$$

When we substitute the stochastic similarity in (6-2) or (6-3) and the kinematical similarity in (6-1) into the above two relations, the scaling P-model is valid only for large and great earthquakes with seismic moments in the range from 1.4×10^{28} dyne·cm to 4.2×10^{29} dyne·cm. Meanwhile, the constant P-model is valid for earthquakes with seismic moments larger than 3.3×10^{26} dyne·cm. This suggests that the constant P-model is preferable to the scaling P-model for large and great earthquakes. This conclusion is confirmed in the next section by analyzing the seismic-wave energy of large and great earthquakes.

6.2. SEISMIC-WAVE ENERGY OF COMPLEX FAULTING PROCESS

The seismic-wave energy radiated from large and great earthquakes provides another constraint on the complex faulting process. Seismic-wave energy is important in the source spectrum over the frequency range $0 \leq \omega \leq \infty$. Seismic moment affects the source spectrum at very low frequency. Earthquake magnitudes depend on the spectral content at the respective characteristic frequencies. Thus, these three quantities represent the source spectrum at different frequencies. Figure 6-1 shows the seismic-wave energy, obtained from recent broad-band analyses of large earthquakes, plotted against seismic moment. The theoretical relationship between S-wave energy and seismic moment from the complex faulting process is also plotted in the figure.

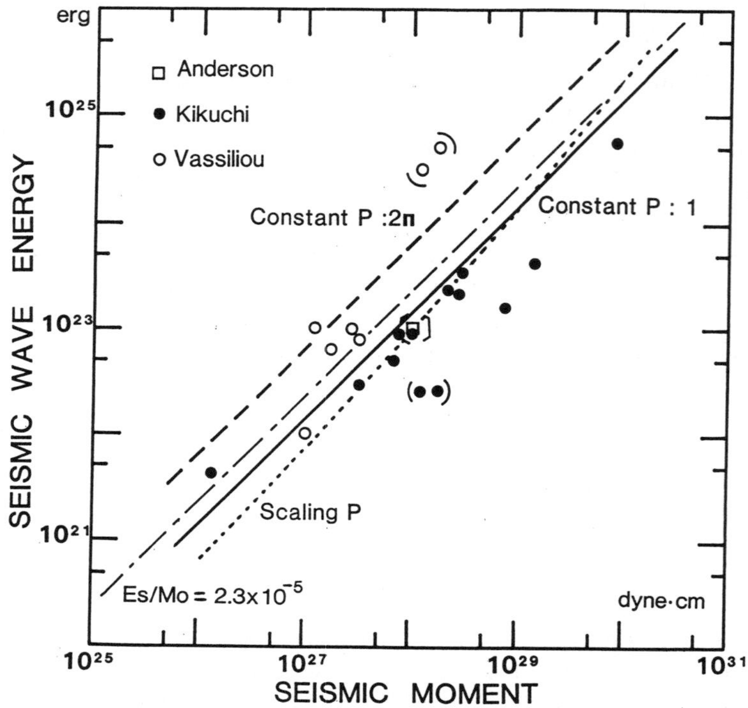

Figure 6 – 1. Seismic-wave energy E_S and seismic moment M_o for large and great earthquakes in subduction zones. The data by solid circles are quoted from Kikuchi and Fukao(1988). Open circles represent data from Vassiliou and Kanamori(1982). E_S values were based on a point source and doubled to be the measure for a finitely propagating fault. The square data point is from strong motion records of the 1985 Michoacan earthquake in Mexico (Anderson *et al.*, 1986). The data in brackets and square brackets are for the same earthquake. Theoretical relationships are indicated in the figure: the constant P-model ($\lambda = 1$) by a solid curve, the constant P-model ($\lambda = 2\pi$) by a dashed curve, the scaling P-model by a dotted curve. Empirical relation using Brune model by Kikuchi and Fukao (1988) is also indicated by a chained curve.

Four such relationships are illustrated in Fig. 6-1; the scaling P-model, the constant P-models assuming $\lambda = 2\pi$ and $\lambda = 1$, and an empirical relation of $E_S = 2.3 \times 10^{-5} M_o$. The last relation is obtained from Brune's model. Although these are based on the similarity parameters derived in the previous section, the constant P-model with $\lambda = 1$ and the scaling P-model better fit the observations than the other two; the constant P-model with $\lambda = 2\pi$ generally overestimates the seismic-wave energy. So that the assumption of $\lambda = 2\pi$ employed in the preceding chapters is not appropriate for large earthquakes. Although a better fit to the data is obtained with

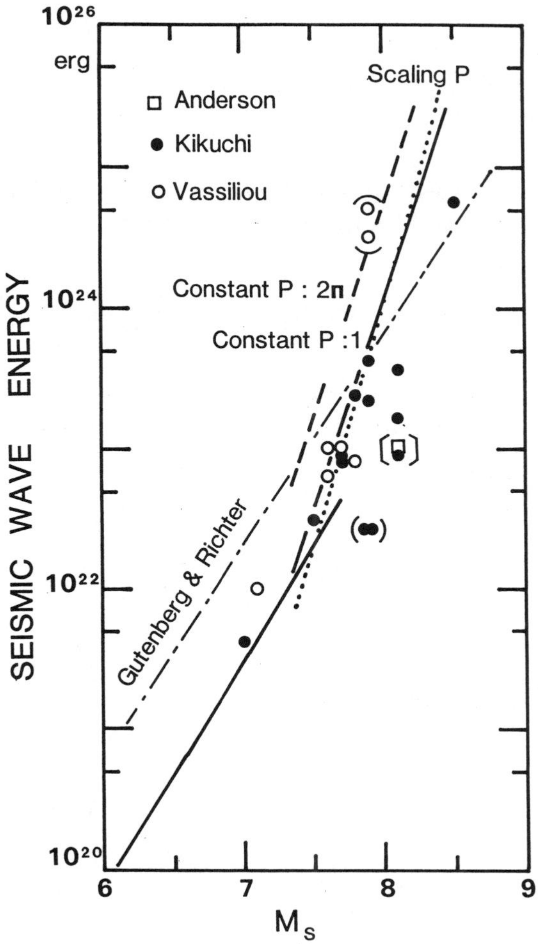

Figure 6 – 2. Seismic-wave energy E_S and surface-wave magnitude M_s for large and great earthquakes in subduction zones (Koyama, 1994). The data and symbols are the same as in Fig. 6-1. In addition, a theoretical relationship for large earthquakes is shown by a solid curve in the magnitude range smaller than 7.5. Gutenberg and Richter's empirical relation is indicated by a chained curve.

a constant patch corner frequency corresponding to $\lambda = 1$ rather than of $\lambda = 2\pi$, all the results in the previous sections are still valid. This is because patch corner frequency of $\lambda = 1$ is also much higher than the characteristic frequency for surface-wave magnitude determination and than the corner frequencies of large earthquakes.

This conclusion for the constant patch corner frequency can be confirmed from the relation between the seismic-wave energy and surface-wave magnitude. A theoretical relationship between seismic-wave energy and surface-wave magnitude for small earthquakes can be derived from (4-16) and (5-3) with (2-15)

$$\log E_S \simeq M_S + \log\Big((1+\frac{\sigma^2}{\bar{a}^2})\bar{a}\frac{\pi\mu\bar{v}^2 p}{4\beta^3}g(\chi)\Big) - \epsilon, \qquad (6-7)$$

where $g(\chi)$ is found in (3-3). Similarly, a relationship for large earthquakes can be obtained from (4-16) and (5-5) with (2-15)

$$\log E_S \simeq \frac{3}{2}M_S + \log\Big((1+\frac{\sigma^2}{\bar{a}^2})\bar{a}^{1/2}\frac{\mu\bar{v}}{8\beta^3 q^{1/2}}g(\chi)\Big) - \frac{3}{2}\epsilon. \qquad (6-8)$$

A theoretical relationship for great earthquakes can be derived from the constant P-model similarly from (4-16), (5-27) and (2-15) as

$$\begin{aligned}\log E_S \quad &\simeq \quad 3M_S + \log(\bar{a}^2+\sigma^2) - \frac{3}{2}\log(\frac{\sigma^2}{\lambda}) \\ &+ \log\Big(\frac{\mu\omega_s^3}{128\beta^3\bar{v}^2}\frac{g(\chi)}{p}\frac{\pi^{3/2}}{(2pq)^{1/2}}f^{-3}(N)\Big) - 3\epsilon, \qquad (6-9)\end{aligned}$$

where ω_s is the characteristic frequency for surface-wave magnitude determination. For the scaling P-model it is from (4-16), (5-26) and (2-15) as

$$\begin{aligned}\log E_S \quad &\simeq \quad 3M_S + \log(\bar{a}^2+\sigma^2) - \frac{3}{2}\log(\lambda\sigma^2) \\ &+ \log\Big(\frac{\mu\omega_s^6}{64\beta^3\bar{v}^2}\frac{g(\chi)}{p}\frac{\pi^{3/2}}{(pq)^{1/2}}f^{-3}(N)\Big) - 3\epsilon. \qquad (6-10)\end{aligned}$$

Since we have determined the similarity parameters, the above relationships describe the $E_S - M_S$ relation for different sized earthquakes. Figure 6-2 shows the relationships between seismic-wave energy and surface-wave magnitude in (6-8), (6-9) and (6-10). The symbols are the same as those in Fig. 6-1; the solid curve indicates a relationship for the constant P-model with $\lambda = 1$, the dashed curve for the constant P-model with $\lambda = 2\pi$, the dotted curve for the scaling P-model, and the chained curve for Gutenberg and Richter's empirical relation in (4-1). The relationship in (6-8) is also illustrated by a solid curve for M_S smaller than 7.5.

We may not be able to retrieve the radiated energies of large earthquakes completely. In spite of that, the scatter of the data is large. Among the theoretical relationships, the relationship for the constant P-model with $\lambda = 1$ satisfies the general trend of the data distribution better than the others. This supports the constant patch corner frequency of the complex

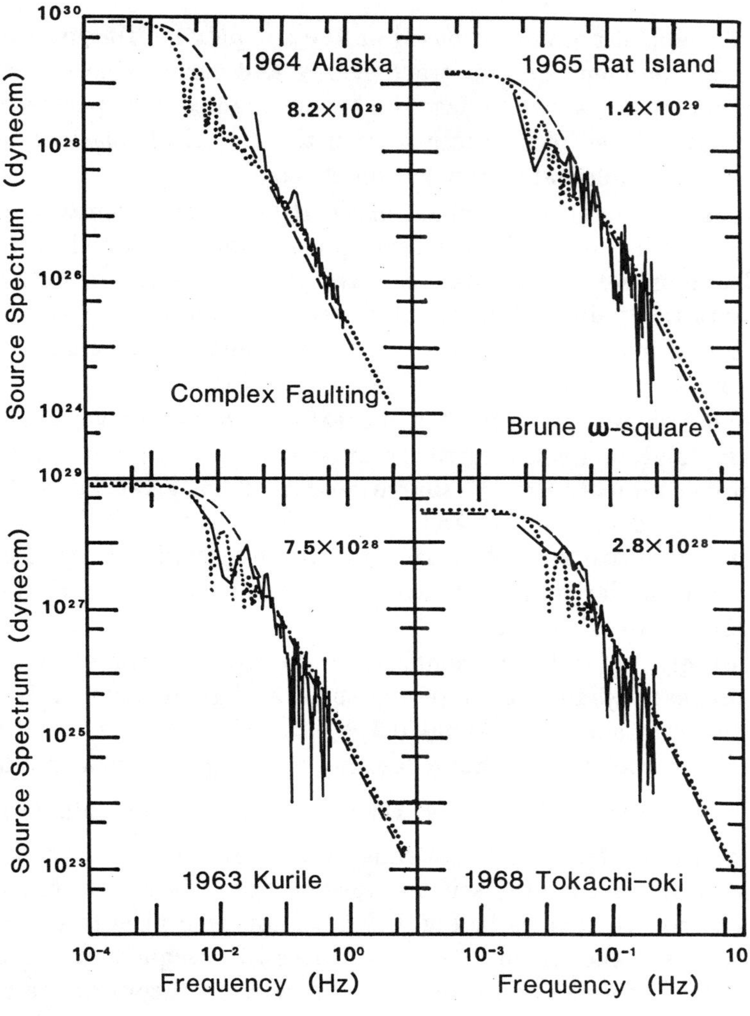

Figure 6 – 3. Displacement source spectra of great earthquakes in subduction zones. The 1964 Alaskan earthquake is reproduced from Houston and Kanamori(1986), the 1965 Rat Island, the 1963 Kurile, and the 1968 Tokachi-oki earthquakes are reproduced from Kikuchi and Fukao(1988). Dotted curves illustrate the source spectrum of the complex faulting process for each earthquake assuming the seismic moment in the figure, with a constant patch corner frequency $\lambda = 1(0.16\text{Hz})$. Dashed curves indicate Brune's spectra.

faulting process for great earthquakes. The assumption of $\lambda = 1$ is crucial for the complex faulting process and only the constant patch corner frequency model satisfies constraints from the independent empirical relations between $M_o - M_S$, $M_o - S$, $M_o - m_b$, $E_S - M_o$ and $E_S - M_S$.

6.3. GENERALIZED SCALING OF EARTHQUAKE SOURCE SPECTRA

We have developed arguments based on the complex faulting process specifying seismic moment, seismic-wave energy and surface-wave magnitude. The parameters necessary to describe the average source spectra of natural earthquakes have been obtained from the empirical relations, and are independent of whole-path propagation effects.

Previous studies have examined observations of the excitation of high-frequency seismic waves of great earthquakes, and concluded that the ω-square Brune model explains the observed source spectra. Since the Brune model is intuitively derived for circular faults, it is not appropriate for large earthquakes which are clearly long and narrow faultings. This has long been an anomaly.

Figure 6-3 shows source spectra of great earthquakes in various subduction zones. Solid curves are retrieved source spectra of great earthquakes. Dashed curves indicate Brune's source spectra for the given value of seismic moment, where $\Delta\sigma_0 = 30$ bar and $\beta = 4$km/s are assumed. Spectra shown by dotted curves are evaluated from the complex faulting process applying stochastic similarity in (6-3) with the constant patch corner frequency $\lambda = 1$, which is about 0.16Hz.

Surprisingly, but not unreasonably, the source spectrum of the complex faulting process explains the observed spectra of great earthquakes over a high frequency range up to about 0.5 $\sim$ 1 Hz. The complex faulting process was derived for a rectangular faulting with a geometrical similarity of $p = \dfrac{W}{L} = 0.5$. Thus it is not necessary to assume an unrealistic faulting mode on a circular fault, such as Brune's model, in order to account for the source spectrum and for the seismic-wave energy of large and great earthquakes. The source spectra in Fig. 6-3 have not been used in constraining the parameters of the complex faulting process. Consequently, this result is further evidence for the applicability of the present theory of the complex faulting process to natural earthquake sources.

The complex faulting process characterizes the displacement source spectrum as a constant spectrum at very low-frequency, with a spectral envelope decaying inversely proportional to ω-square above the corner frequency and decaying less rapidly in and around the patch corner frequency. Although similar source spectra has been proposed for large earthquakes, for example the revised model B by Aki (1972), the previous studies analyzed empirical relations of seismic moment and different magnitude scales to provide constraints on the relative spectral content. The present theory of the complex faulting process was based on the five constraints derived from the observations and the theory generally describes the spectral content at high frequencies as well as the seismic-wave energy of large and great earthquakes.

Note that the complex faulting process specified by the constant patch corner frequency of $\lambda = 1$ is not valid for earthquakes with surface-wave magnitude M_s smaller than 7. This is because the assumed patch corner frequency approaches the corner frequency of moderate-size earthquakes and the criterion in (6-5) is no longer valid. Therefore, the present theory of the complex faulting process can be applied only to subduction zone earthquakes with magnitude larger than 7.5.

6.4. SEISMIC EFFICIENCY OF COMPLEX FAULTING PROCESS

The energy budget of large earthquakes is considered in this section to conclude the study of the complex faulting process. Since the complexity provides prodigious sources of static and kinetic energies, we investigate the role of the fault heterogeneities on the seismic energy. The static estimate of seismic energy in (4-12) can formally be rewritten as

$$W_h = \frac{\Delta\sigma_0}{2\mu} M_o \Big\{ 1 + \frac{\sigma^2}{\bar{a}^2} \Big(\frac{2}{\lambda T_0} \Big)^{1/2} \Big\}. \qquad (6-11)$$

Since $\dfrac{\bar{\sigma}_1 + \bar{\sigma}_2}{\Delta\sigma_0} = 1$ in the above, this estimate corresponds to the heterogeneous faulting with 100% stress drop. The seismic-wave energy in (4-17) is approximately expressed as

$$E_S \simeq \frac{\Delta\sigma_0}{2\mu} M_o \Big\{ 1 + \frac{\sigma^2}{\bar{a}^2} \Big\} \frac{p}{4q} \chi^3 g(\chi), \qquad (6-12)$$

where $g(\chi)$ is a function of the velocity ratio $\chi = \dfrac{\bar{v}}{\beta}$ defined in (3-3). Thus we obtain a relation for the ratio of seismic-wave energy to moment:

$$\frac{E_S}{M_o} = \frac{\Delta\sigma_0}{2\mu} \times 0.229 \Big\{ 1 + \frac{\sigma^2}{\bar{a}^2} \Big\}, \qquad (6-13)$$

where $\chi = 0.8$, and $p = q = \dfrac{1}{2}$ are assumed. By substituting the stochastic similarity and the average dislocation velocity in (6-13), the above relation is consistent with the empirical relation $E_S = 2.3 \times 10^{-5} M_o$ derived from Brune's model. Note that the seismic-wave energy of the stochastic part is about 1.5 times larger than that of the deterministic part of the complex faulting process, when the constant patch corner frequency of $\lambda = 1$, with (6-1) and (6-3).

The ratio of $\dfrac{E_S}{W_h}$ is the seismic efficiency η_h for the case of 100% stress drop. Evaluated from (6-11) and (6-12), it is about 0.47 for a great earthquake with $M_S = 8.5$ and about 0.32 for $M_S = 7.5$. The values become much smaller when frictional stress is considered. This kind of variation in seismic efficiency has been pointed out for natural earthquakes. Since the seismic efficiency of previous models is independent of the earthquake source size, this variation in seismic efficiency is considered to result from the energy loss due to cohesion which is restricted to the rupture front and increases in proportion to the length of the fault, whereas the seismic-wave energy increases in proportion to the area of the fault. Therefore, seismic efficiency is considered to be larger for great earthquakes than for large earthquakes, based on the previous models without heterogeneities.

In contrast to the above argument, the complex faulting process proposes another interpretation. The contribution of fault heterogeneities to the total work W_h is smaller for larger earthquakes than for smaller events: Theoretically, this can be found in the second term in braces of (6-11), where the parameter T_0 increases with the size of earthquake, reducing the contribution of fault heterogeneities. Consequently, the energy ratio of $\frac{E_S}{W_h}$ increases in case of the complex faulting process of great earthquakes without considering the size dependency of frictional energy loss. The variation in $\frac{E_S}{W_h}$ is fundamental to the size effect of the deterministic and stochastic faulting processes and is not attributed merely to the frictional and cohesive conditions on the fault.

CHAPTER 7

STOCHASTIC RUPTURE PROCESS OF FAULT PATCHES

Fluctuation phenomena with $1/f$ spectrum have been observed in a wide variety of dissimilar physical systems, where f is frequency. Noise experiments in electronic devices reveal the fractional power spectrum of $1/f^{\alpha}$ type. It is known that noise currents in semiconductors and turbulent flow fields are described by fractional power spectra of $1/f^{3/2}$ and $1/f^{5/3}$ types. We have shown that the acceleration spectrum is characterized by a fractional power in §3, the dynamical process of random activation of fault patches is now examined in relation to $1/f^{\alpha}$ fluctuations.

Various theories have been proposed to explain such $1/f$ spectral behavior and its relation to the long-tail behavior of complex systems. Most theories start with an a priori assumption for the autocorrelation function of a particular process and/or a probability function for the random events. There is little discussion on what conditions and material properties are necessary for the functions and what is the physical significance of the functional forms. The analysis here is not aimed at obtaining $1/f^{\alpha}$ spectrum by assuming an autocovariance function nor the stochastic property of the respective process. We will study the fundamental property of the time evolution of the random activation of fault patches in a general manner.

The excitation of short waves by large earthquakes is strictly related to the random fracturing of small-scale heterogeneous areas where the stress drops are not constant but fluctuating. Fracturing one heterogeneous area of fault patches radiates one pulse of oscillations. Fracturing many fault patches in a random manner generates a random-pulse time series with incoherent phases. Therefore, the rupture process of random fault patches of the complex faulting process is observed as a random-pulse time series. This leads us to consider that the rupture process of random fault patches could be simulated by a stochastic process of rupture propagation on small-scale heterogeneous areas activated in a random fashion.

7.1. RUPTURE PROCESS OF RANDOM FAULT PATCHES

Generally speaking, the dynamics of random systems is so complicated that it is difficult to characterize the system by only its macroscopic parameters. The elementary process of the system may be unknown nor do we know the dynamic temporal evolution of the elementary process. Let us consider here a fault plane with heterogeneities of all scales distributed over the plane. Rupturing a particular fault patch radiates one displacement pulse. Rupturing many of these fault patches in a random manner generates random pulses with incoherent phase. Since we consider a variety of scales for random fault patches on the fault plane, there are many different responses due to these ruptures. Within the distribution of fault patches, there will be numbers which are characterized by the same response function. We classify these fault patches into the same category. The whole system of complex faulting process is composed of a cluster of many categories of elementary events. Each category of events is represented by a set of elements in which the respective stochastic behavior is governed by the same equation. This means that the state of the whole system is expressed as a sum of *local states* of categories of events: Let the state of the system be $X(t)$, and of the local states $X_j(t)$, then we have

$$X(t) = \sum_{j=0} X_j(t). \tag{7 - 1}$$

Suppose that the state of the system $X(t)$ is described by

$$\dot{X}(t) + \bar{\gamma} X(t) = n(t), \tag{7 - 2}$$

where the dot stands for the time derivative, $\bar{\gamma}$ is a positive constant, and $n(t)$ represents a random force.

If the whole system is expressed by a single category of events $X_0(t)$, equation (7-2) is reduced to:

$$\dot{X}_0(t) + \bar{\gamma} X_0(t) = n_0(t). \tag{7 - 3}$$

The random force in this case is assumed to have zero mean

$$< n_0(t) >= 0, \tag{7 - 4}$$

and the autocovariance function

$$< n_0(t+\tau) n_0(t) >= \varrho^2 \delta(\tau), \tag{7 - 5}$$

where ϱ^2 is a constant depending upon the frequency of random activation in unit time, and δ is Dirac delta function. Mean values are denoted by brackets $<>$.

Clearly, equation (7-2) is too simplified to represent generally the complex system. However, the primary interest in this chapter is to describe the stochastic behavior of the complex system, and not to derive a model appropriate to the actual detail. Note that the simplified equation (7-3) provides its spectrum with a characteristic corner frequency of $\bar{\gamma}$. It is well known that equation (7-2) is the Langevin equation to describe classical Brownian motion.

Subscript 0 in (7-3) indicates a generator of the system. The generator is composed of elements activated by $n_0(t)$ with a characteristic decay of $\bar{\gamma}$.

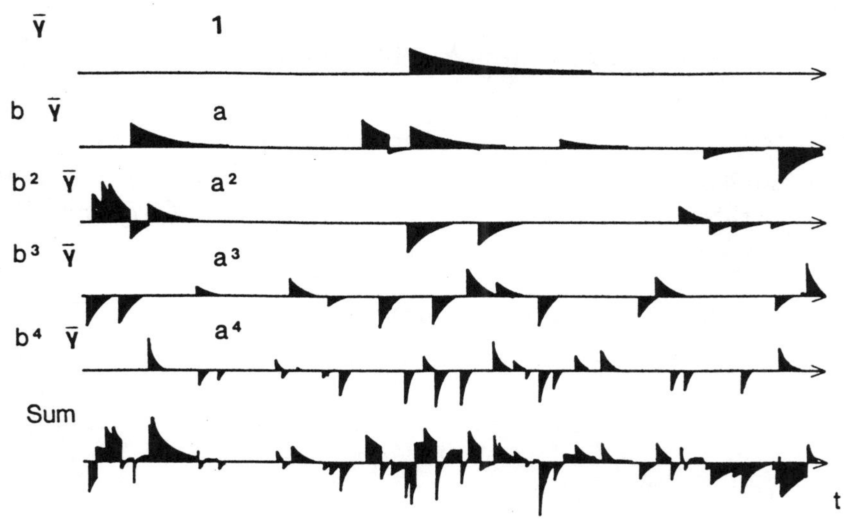

Figure 7 – 1. **Response of the complex system to random activation. Amplitude and unit of time in the abscissa are arbitrary. Self-similar sets of categories of events are illustrated in rows. Each category is composed of random-activated pulses with the same decay-rate but with random amplitude.**

$< X_0(t) >$ becomes zero as $t \rightarrow \infty$, because of (7-4). The autocovariance function $C_0(\tau)$ of $X_0(t)$ is calculated as

$$\begin{aligned} C_0(\tau) &= < X_0(t+\tau)X_0(t) > \\ &= \frac{\varrho^2}{2\bar{\gamma}} \exp(-\bar{\gamma}|\tau|), \end{aligned} \qquad (7-6)$$

under the initial condition $X_0(-\infty) = 0$.

7.1.1. *Complexity of Random Fault Patches*

There are many categories of events due to random fault patches, in which each stochastic behavior is governed by an equation as in (7-2). Therefore, the whole response of the rupture process of all random fault patches can be expressed by the sum of responses of local states. Now we introduce a scaling rule to model the complexity of the whole response of the complex rupture process. Scaling parameters a and b are considered to scale local states of X_{j-1} and X_j, and also scale n_{j-1} and n_j $(j = 1, 2, ..., N)$, where N is the number of local states of categories. Since X_j and n_j are of random functions and are statistically independent, the scaling rule is not to specify the self similarity and/or self affinity of the functional forms but to describe the similarity nature underlying the statistical properties of X_j and n_j:

$$f_X \quad : \quad \sqrt{\frac{a}{b}}\, X_{j-1}(bt) \to X_j(t), \tag{7 - 7}$$

$$f_n \quad : \quad \sqrt{ab}\; n_{j-1}(bt) \to n_j(t), \tag{7 - 8}$$

where a and b are positive real, and subscript j indicates the j-th local state after the j-th scaling. Consequently, the scaling rule in (7-7) and (7-8) is *stochastic.* We have from (7-3), (7-7), and (7-8)

$$\begin{aligned} \dot{X}_1(t) + b\bar{\gamma}X_1(t) &= n_1(t), \\ \dot{X}_2(t) + b^2\bar{\gamma}X_2(t) &= n_2(t), \end{aligned} \tag{7 - 9}$$

and so on. The autocovariance of random force is expressed similarly

$$\langle n_j(t)n_j(t+\tau)\rangle = a\langle n_{j-1}(t)n_{j-1}(t+\tau)\rangle \quad (j = 1, 2, ..., N). \tag{7 - 10}$$

The autocovariance function of each local state is then scaled as

$$C_j(\tau) = \frac{a}{b}\, C_{j-1}(b\tau). \tag{7 - 11}$$

Although we introduced the stochastic scaling rule in (7-7) and (7-8) as mapping functions of X_{j-1} and n_{j-1} to X_j and n_j, the scaling relation of the random force and the autocovariance is defined in (7-10) and (7-11). For pulses generated by (7-9), each level of the scaling specifies a response of a local state with b times more rapid decay than that of the younger scaling level. The responses are triggered a times as often in a random manner, so that this is a measure of the activation rate. Different activation rates for different local states of categories is a manifestation of many threshold levels for the activation of respective elements.

7.1.2. *Orthogonality of Random Fault Patches*

An impulse response $h_j(t)$ of the j-th category of events can be obtained from (7-9) as

$$h_j(t) = \exp(-b^j\bar{\gamma}t)U(t), \tag{7 - 12}$$

where $U(t)$ is the unit step function. The response of the j-th local state can be expressed by convolution of the impulse response $h_j(t)$ and the random activator $n_j(t)$ as

$$X_j(t) = \int_{-\infty}^{t} n_j(s)h_j(t-s)ds. \tag{7 - 13}$$

Cross correlation between j-th and k-th local states can be calculated

$$\begin{aligned} R_{X_j X_k}(t_1, t_2) &= \langle X_j(t_1) X_k(t_2) \rangle \\ &= \int_{-\infty}^{t_1} du \int_{-\infty}^{t_2} ds \langle n_j(u) n_k(s) \rangle h_j(t_1 - u) h_k(t_2 - s). \end{aligned} \tag{7 - 14}$$

Since $n_j(u)$ and $n_k(s)$ are random forces, they are orthogonal each other:

$$\langle n_j(u) n_k(s) \rangle = \begin{cases} a^j \varrho^2 \delta(u - s) & (j = k), \\ 0 & (j \neq k). \end{cases} \tag{7 - 15}$$

Therefore the local states of random fault patches are orthogonal. This property yields a set of orthogonal function derived from the scaled Langevin equation for the responses of rupture process of random fault patches.

The autocovariance function of the whole response is, then, expressed by the sum of (7-11) as

$$C(\tau) = \frac{\varrho^2}{2\bar{\gamma}} \sum_{j=0}^{N} \left(\frac{a}{b}\right)^j \exp(-b^j \bar{\gamma} |\tau|). \tag{7 - 16}$$

The parameter $\dfrac{\varrho^2 a^j}{2\bar{\gamma} b^j}$ can be used to normalize the j-th local state to define an orthonormal system of functions. The scaling level from $j = 0$ to N specifies a set of impulse responses for random fault patches in (7-12). The impulse response has been expressed by a characteristic (angular) frequency of $b^j \bar{\gamma}$ for the j-th category of events. This characteristic frequency covers a range from $\bar{\gamma}$ to $b^N \bar{\gamma}$. This defines the scaling region of random fault patches, studied in §3.

Figure 7-1 is a sketch of the random responses activated in this context. When $\dfrac{a}{b} > 1$, the term of $\left(\dfrac{a}{b}\right)^j$ within the summation in the right hand side of (7-16) diverges whereas the exponential function converges. We apply the steepest descent method for (7-16) considering large $|\tau|$. The approximation is briefly described in Appendix G to obtain,

$$C(\tau) \simeq A_\kappa \, |\tau|^{-(\kappa - 1)}, \tag{7 - 17}$$

$$A_\kappa = \sqrt{\frac{2\pi}{\kappa - 1}} \frac{\exp([\kappa - 1][\ln(\kappa - 1) - 1])}{\ln(b)} \frac{\varrho^2}{2},$$

where κ is a fractal dimension of the random activation defined by the scaling parameters a and b as

$$\kappa = \frac{\ln(a)}{\ln(b)}. \qquad (7-18)$$

Here $\bar{\gamma} = 1$ has been assumed in order to normalize time and frequency without loss of generality. Note that $\kappa > 1$ in this case because $\frac{a}{b} > 1$. The parameter b then specifies the amplitude decay of responses as found in Fig. 7-1. The parameters a and b are left free here, but will be determined latter.

Long-tail behavior of the complex system is shown by (7-17). The behavior is represented by the power law decay in time as in (7-17) not by exponential decay as in (7-6); this is commonly found for natural phenomena and disordered materials and has been studied extensively as complex systems. However, we can learn more beyond the general trend and relative behavior, since equation (7-17) is the solution of the Langevin equation without any a priori assumption of the probability density and/or the autocovariance function of the system.

7.2. FRACTIONAL POWER SPECTRUM OF THE COMPLEX SYSTEM

The power spectrum of the system is calculated from the autocovariance in (7-17)

$$\begin{aligned} P_s(\omega) &= \int_{-\infty}^{\infty} C(\tau)\exp(-i\omega\tau)d\tau \\ &= 2A_\kappa\Gamma(2-\kappa)\ \cos(\frac{\pi}{2}[2-\kappa])\ |\omega|^{\kappa-2} \\ &\quad (2-\kappa \neq integer), \end{aligned} \qquad (7-19)$$

where Γ is the Γ-function, and ω is angular frequency and $\bar{\gamma}$ is taken to be the normalization factor for t and ω. Equation (7-19) is the integral representation of the Γ-function as a hyperfunction. This describes fractional power spectra from a white spectrum, for which κ tends to 2, to $1/f$ spectrum, for which κ tends to 1.

The fractional power spectrum in (7-19) does not always hold for the entire frequency range because of the assumption in deriving (7-17). The power spectrum is white at very low frequencies $\omega \ll 1$ because the spectrum in the low frequency limit represents that of random impulses and not of random pulses with finite time durations. For extremely high frequencies, $\omega > b^N$ (N being the maximum scaling level), the spectrum is $1/\omega^2$ type, since each $C_j(\tau)$ in (7-6) and (7-11) is attributed to the corresponding Lorentz spectrum with $1/\omega^2$ decay.

Although the integral formula in (7-19) is valid for $2 < \kappa < 3$, we observe a negative power spectrum because of the negative value of Γ function. This is physically unrealistic. The reason is that integral formula applied in (7-19) used the integral representation of the Γ-function. The Γ-function in this case is evaluated as the finite part of hyperfunctions excluding singularities. Consequently, the power spectrum in (7-19) is valid for $1 < \kappa < 2$ and not for $2 < \kappa < 3$. In the latter case, we need a different approach for evaluating the power spectrum.

The power spectrum of each local state is evaluated from (7-16). The spectrum of the whole system is expressed by the sum of the power spectra of local states, because of the orthogonality

$$P(\omega) = \sum_{j=0}^{N} \left(\frac{a}{b}\right)^j \frac{\varrho^2 b^j}{b^{2j} + \omega^2}. \qquad (7-20)$$

The power spectrum of j-th local state is described by the Lorentz spectrum with a characteristic frequency of b^j. Note that the value has been normalized by $\bar{\gamma}$. In the low frequency range $\omega < b^j$, it is a white spectrum, whereas in the high frequency range $\omega \gg b^j$, the spectrum shows a frequency dependence of $1/\omega^2$. Since the characteristic frequencies cover the range from 1 to b^N with the scaling level of j, the random system is considered within this scaling region of characteristic frequencies.

Figure 7-2 shows example power spectrum described by (7-20). On the left, are Cantor bars for the scaling parameters $a = 2$ and $b = 3$. The fractal dimension $\kappa_C = \dfrac{\ln 2}{\ln 3}$ is about 0.63. On the right, are Koch curves for $a = 4$ and $b = 3$. The fractal dimension $\kappa_K = \dfrac{\ln 4}{\ln 3}$ is about 1.26. Two corner frequencies are found for each spectrum in Fig. 7-2: One is $\omega = 1$, corresponding to that of the generator in the case $j = 0$. The other is the largest characteristic frequency of b^N for the $j = N$-th local state. A constant power spectrum is found for the frequency range lower than the scaling region and ω^{-2} spectral behavior at higher frequencies. The fractional decay of power spectra is dependent on the fractal dimensions of κ_C and κ_K and is confined to the scaling region of frequency. Measured spectral decay in Fig. 7-2(a) is about 1.3 and is about 0.74 in (b). These values agree with the theoretical $2 - \kappa$ from the asymptotic form of (7-19).

Now we consider a scaling of $\omega \to \dfrac{\omega}{b}$ in (7-20). The power spectrum is then expressed by a scaling relation of

$$P\left(\frac{\omega}{b}\right) = \frac{b^2}{a}\left\{P(\omega) - \frac{\varrho^2}{1 + \omega^2} + \frac{\varrho^2 a^{N+1}}{b^{2(N+1)} + \omega^2}\right\}. \qquad (7-21)$$

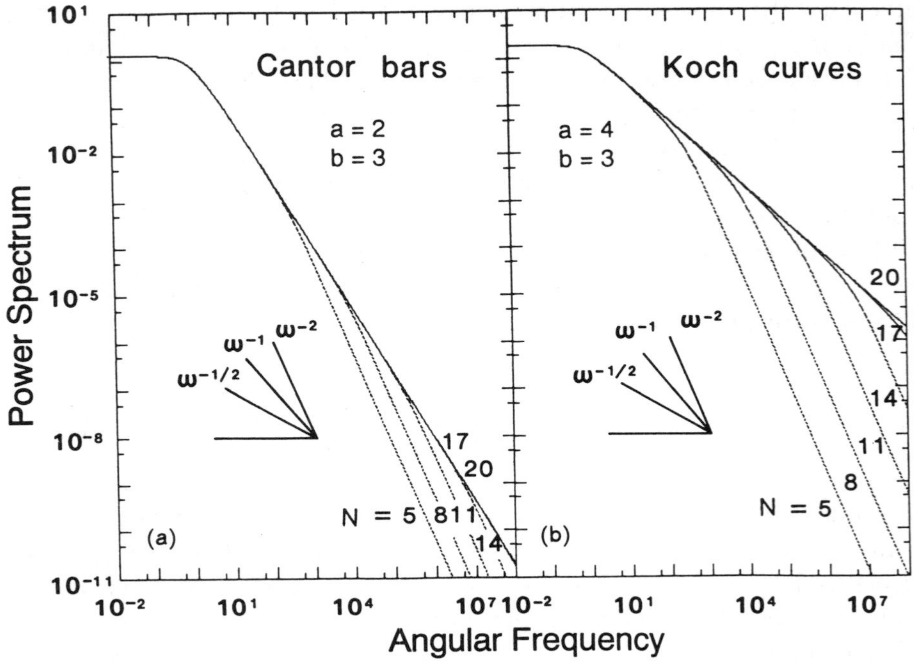

Figure 7 – 2. Fractional power spectrum of the complex system due to random activation of small-scale fault patches. Scaling parameters a and b are assumed to construct self similar fault patches (a) characterized by Cantor bars and (b) by Koch curves. Scaling level N is taken as a parameter, specifying the scaling region of frequency. Spectral decay between two corner frequencies of the scaling region is about 1.3 in (a) and about 0.74 in (b). The spectral decays are consistent with the theory of the asymptotic form in (7-19).

This is not a simple scaling relation, but an asymptotic form can be evaluated as follows. When $a < b^2$, the scaling relation (7-21) is

$$P(\omega) \simeq \frac{a}{b^2} \; P(\frac{\omega}{b}), \tag{7 - 22}$$

for the frequency range within the scaling region $1 \ll \omega \ll b^{N+1}$. When $a > b^2$, the first and second terms in the right-hand side of (7-21) become small compared to the third after repeated scaling. Then, equation (7-21) is expressed approximately as

$$P_e(\omega) \simeq \left(\frac{a}{b}\right)^N \; \frac{\varrho^2 b^N}{b^{2N} + \omega^2}. \tag{7 - 23}$$

This is the case $\kappa > 2$ since $a > b^2$. Spectrum (7-23) is a Lorentz spectrum with a characteristic corner frequency of b^N.

Another approximation for evaluating the integral in (7-20) can be obtained by the Euler-Maclaurin formula as

$$P_e(\omega) \simeq \int_0^N \left(\frac{a}{b}\right)^x \frac{\varrho^2 b^x}{b^{2x}+\omega^2}dx - \sum_{j=0}^{N-1}\int_j^{j+1}\left\{\left(\frac{a}{b}\right)^x \frac{\varrho^2 b^x}{b^{2x}+\omega^2}\right\}' dx, \quad (7-24)$$

where $'$ stands for the derivative. In the above, we relax the restriction for the fractal dimension to $\kappa > 0$.

The approximation (7-24) is rewritten taking a new variable of $t = |\omega|b^x$ as

$$P_e(\omega) \simeq \frac{\varrho^2}{\ln(b)}|\omega|^{\kappa-2}\int_{1/|\omega|}^{b^N/|\omega|}\frac{t^{\kappa-1}}{t^2+1}dt - \sum_{j=0}^{N-1}\int_j^{j+1}\left\{\left(\frac{a}{b}\right)^x \frac{\varrho^2 b^x}{b^{2x}+\omega^2}\right\}' dx. \quad (7-25)$$

Since our interest is confined to the power spectrum in the scaling region of frequency $1 \ll \omega \ll b^N$, the first integral can be expanded from 0 to ∞. Then we apply the branch cut integral to evaluate (7-25) for $0 < \kappa < 1$ as

$$P_e(\omega) \simeq \frac{\varrho^2\pi}{\ln(b)}\frac{\sin(\frac{\pi}{2}[\kappa-1])}{\sin(\pi[\kappa-1])}|\omega|^{\kappa-2} + 0(\frac{\varrho^2}{1+\omega^2}). \quad (7-26)$$

This spectrum in the limit $\kappa \to 0_+$ is identical to one which we can evaluate by integrating each component of the Taylor series for the numerator in (7-25) as

$$P_e(\omega) \simeq \frac{\varrho^2}{\kappa\ln(b)}|\omega|^{\kappa-2} + 0(\frac{\varrho^2}{1+\omega^2}). \quad (7-27)$$

Similarly when $\kappa \to 2_-$, the power spectrum in (7-26) is reduced to

$$P_e(\omega) \simeq \frac{\varrho^2}{(2-\kappa)\ln(b)}|\omega|^{\kappa-2} + 0(\frac{\varrho^2}{1+\omega^2}). \quad (7-28)$$

This is identical to the power spectrum derived by the steepest descent method in (7-19). The fractional power spectrum obtained in (7-26) to (7-28) behaves as $1/\omega^2$ for $\kappa \to 0$ and as $1/\omega$ for $\kappa \to 1$.

Consequently, these spectra show a fractional power spectrum from $1/f$ to the Lorentz decay. With the result of (7-19), the present complex system generally describes a fractional power spectrum $1/f^\alpha$ from a white to a Lorentzian. The power exponent α of $1/f^\alpha$ spectrum can be expressed as a function of the fractal dimension of random activation as

$$\alpha = \begin{cases} 2-\kappa, & (0 < \kappa < 2); \\ 0 \text{ (or } 2,) & (\kappa > 2). \end{cases} \quad (7-29)$$

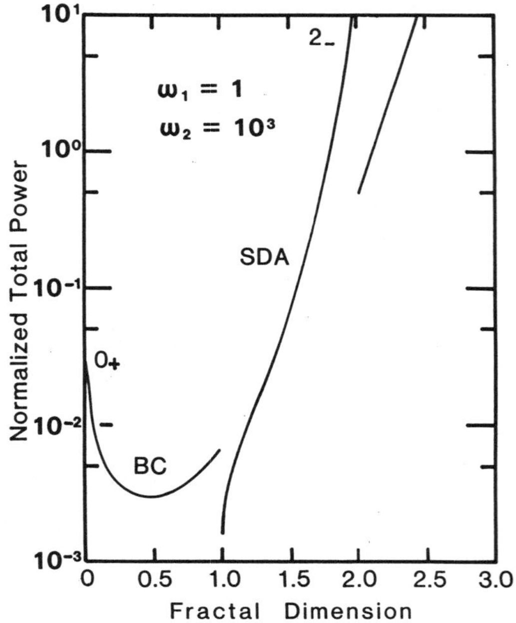

Figure 7 – 3. Total power of the fractional power spectrum as a function of fractal dimension of the scaling parameters (Koyama and Hara, 1992). Maximum and minimum frequencies specify the scaling region of the system response, which are assumed as 1000 and 1. The value of total power is normalized by the maximum frequency. **SDA** indicates the steepest descent approximation in the text, and **BC** the branch-cut integral for the fractal dimension of $0 < \kappa < 1$. $\kappa \to 0_+$ and $\kappa \to 2_-$ are corresponding limiting cases.

This fractional decay of power spectra has been confirmed by the numerical analysis shown in Fig. 7-2.

7.3. TOTAL POWER OF FRACTIONAL POWER SPECTRUM

The total power of the spectrum for the complex system increases monotonically with κ because the number of events included within unit time increases with scaling parameter a. But we do not have a wide interest in the spectrum over the entire frequency band. For frequencies lower than ω_1 (=1; normalized by $\bar{\gamma}$) the power spectrum is always white, above $\omega_2(= b^N)$ it is characterized by $1/\omega^2$ of a Lorentzian. In the scaling region of frequency from ω_1 to ω_2, the power spectrum is characterized by a fractional power decay.

Thus we have the band-limited total power of the fractional power spectrum

$$T_p(\kappa) = \frac{1}{\pi} \int_{\omega_1}^{\omega_2} P_{s,e}(\omega)\, d\omega, \tag{7-30}$$

where $P_{s,e}$ can be found from (7-19) to (7-28). From Parseval's theorem this is equivalent to the band-limited system-energy in unit time. Since the dynamical process of the present system is stationary, we obtain the band-limited total energy by multiplying (7-30) by the time duration of the system.

Figure 7-3 shows T_p as a function of fractal dimension κ. Although κ has been an arbitrary parameter for specifying the dynamics of the complex system, we observe three values of κ, 2, 1 and about 0.47 for which the present complex system is specified by the minimum total power (energy)

$$\frac{dT_p(\kappa)}{d\kappa} = 0. \tag{7-31}$$

When $\kappa > 2$, the responses of many events overlap in the time domain, since $a > b^2$. Such a system is characterized by random-and-dense phenomena, and may not always be stable. The Lorentz spectrum is stable, $\kappa = 2$, and the system is characterized by the local minimum in the band-limited total power. Figure 7-3 also indicates that $1/f$ spectrum is another mode, when $\kappa = 1$, and the scaling parameters satisfy the condition $a = b$. The other steady state is specified by the power spectrum of $1/f^{1.53}$ type. Since $a < b$ in this case, random activation is sporadic in the time domain compared with the characteristic response-duration.

For thermal equilibrium, the steady-state system is characterized by the minimum free energy. In contrast, steady-state propagation of elastic fractures is specified by the minimum potential energy (elastic energy), and the total power $T_p(\kappa)$ in (7-30) corresponds to this elastic energy in the scaling region. Therefore, the above three modes indicate the steady states of the system.

The greater the difference of the power exponent α from 1, the more removed is the physical system from the steady state. This is why $1/f$ spectrum is so ubiquitous in a variety of physical phenomena. This conclusion is derived from the minimum total power criterion by describing the complex system in terms of the fundamental Langevin equation.

Also we have found here that there are two other stable modes specified by the minimum total power criterion. One mode gives the Lorentz spectrum, which is a characteristic of the classical Brownian motion. The other mode is characterized by random-but-sporadic phenomena with a power spectrum close to $1/f^{1.53}$ type, in contrast to the $1/f$ and Lorentzian. Since the system generally describes sparse phenomena in this last case, this mode is one representation for describing intermittence in turbulent flow fields. Since this mode corresponds to the broad minimum in Fig. 7-3, the spectral behavior of $1/f^{1.53}$ type would be consistent with the $\frac{5}{3}$ exponent energy-spectrum of turbulent velocity fields and may be also with the $\frac{3}{2}$ exponent power-spectrum of semiconductor noises. The spectrum characterized by a power exponent of $\frac{5}{3}$ is termed the Kolmogorov spectrum.

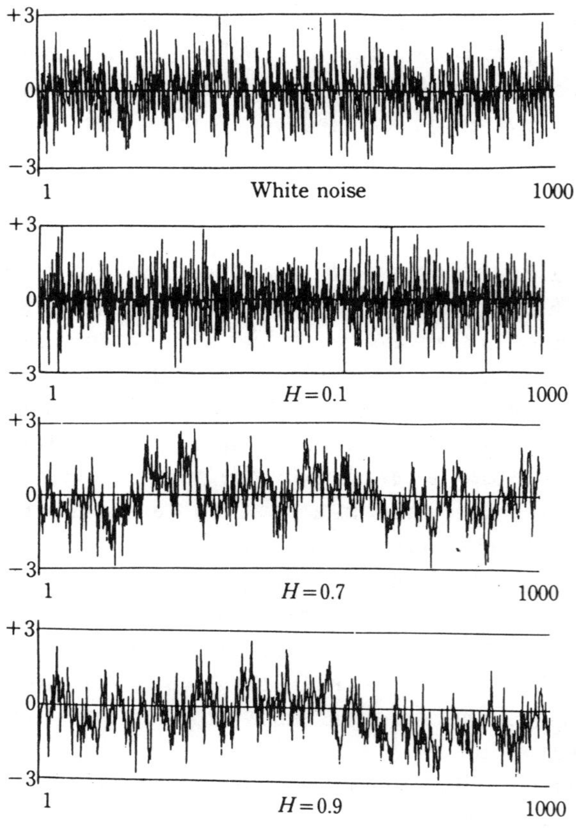

Figure 7 – 4. White noise and fractional Gaussian noise (Mandelbrot and Van Ness, 1968). Vertical axis is arbitrary corresponding to the variance of random noise. Horizontal axis is in time steps up to 1000. H is the Hurst exponent.

7.4. FRACTIONAL BROWNIAN MOTION AND RANDOM FAULT PATCHES

It is well known that the Langevin equation in (7-2) describes the stochastic behavior of classical Brownian motion. Brownian motion is a highly correlated stochastic process, and its derivative may be Gaussian random noise. Gaussian random noise is completely uncorrelated. There is defined a stochastic process, fractional Brownian motion, with temporal and spectral behavior between those of Brownian motion and Gaussian random noise. Fractional Brownian motion is formally obtained by integration of a moving average of Gaussian random noise weighted by a kernel of past time, $t^{H-1/2}$. For $H = \frac{1}{2}$, the process becomes an integration of Gaussian noise and provides the classical Brownian motion.

Figure 7-4 shows Gaussian random noise (white noise) and fractional Gaussian noise obtained by this moving average with $t^{H-1/2}$. In case $H = 0.1$, high frequency components dominate compared to white noise, whereas the case $H = 0.9$ yields dominant long-period components. The latter gives an autocovariance with a large time lag, leading to the long-tail behavior of fractional Brownian motion. Fractional Gaussian noise may be formally considered as the time derivative of fractional Brownian motion similarly to the formal relation between Gaussian noise and classical Brownian motion. The power coefficient in terms of H is very important in characterizing fractional Brownian motion, although H is an empirical parameter ($0 < H < 1$), called the Hurst exponent, and its physical significance is not obvious.

Since the theory in the preceding sections describes the stochastic behavior of complex system based on the Langevin equation with the aid of scaling, the theory gives an insight into fractional Brownian motion from the very basic equation. Figure 7-5 shows the whole response of the present complex system and those of local states derived from the scaled Langevin equation. Since fractional Brownian motion in Fig. 7-5 would be the velocity response and not the trace of Brownian motion, it may be formally equivalent to the fractional Gaussian noise. We find that the general property of the random behavior in Fig. 7-5 agrees quite well with that in Fig. 7-4. This proves the validity of the scaled Langevin formalism for fractional Brownian motion which has been heuristically defined. The primary result in this chapter is the description of the spectral behavior of $1/f^{\alpha}$, where $\alpha = 2 - \kappa$ in (7-29). Fractional Brownian motion produces a power spectrum of the form $1/f^{2H-1}$, where f is frequency. Then we formally obtain a relation between the fractal dimension κ of the scaling parameters and Hurst exponent H as

$$\kappa = 3 - 2H. \qquad (7-32)$$

It is clear that classical Brownian motion $H = \frac{1}{2}$ corresponds to $\kappa = 2$ in the present representation, where the trace of the Brownian motion covers

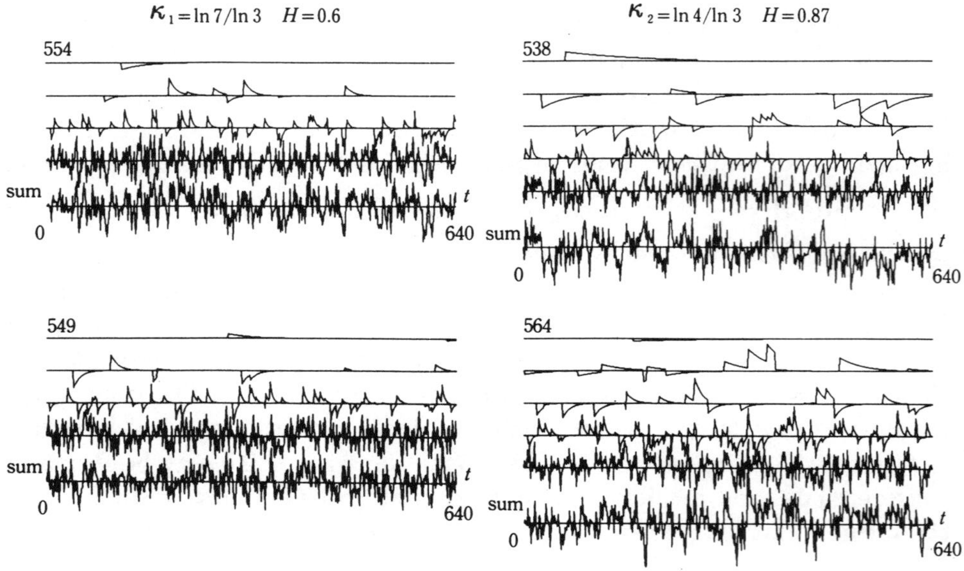

Figure 7 – 5. **Fractional Brownian motion derived from the scaled Langevin equation. Fractal dimension of the scaling is indicated by κ, which relates to Hurst exponent as $\kappa = 3 - 2H$. Two examples for each value of κ_1 and κ_2 are simulated. Unit of vertical axis is arbitrary depending on the variance of random motions. Abscissa units are discrete time steps up to 640. The response functions of the first five local states and the response of the whole system are shown (The first local state is empty for κ_1 cases).**

the whole two-dimensional space. For $\frac{1}{2} < H < 1$, the process is characterized specifically by the long-tail behavior, where $1 < \kappa < 2$. This indicates that the motion is characterized by self-avoiding Brownian motion, in which a Brownian particle does not return to the origin. Infinitely long-tail behavior of the system gives $1/f$ spectrum when $H \to 1_-$ and $\kappa \to 1_+$.

The fractal dimension becomes $2 < \kappa < 3$, when $0 < H < \frac{1}{2}$. The trace of Brownian motion fills ordinary three-dimensional space in this case. On a two-dimensional plane, the motion shows an unbounded trail which visits any prescribed point infinitely often. All these show that the degree of correlation of fractional Brownian motion by H is represented by the fractal dimension κ of the present complex system. We also understand the geometrical property of fractional Brownian motion in terms of H and κ†.

†The trace of fractional Brownian motion has a spectral density proportional to $1/f^{2H+1}$, where f is frequency. The spectral behavior here may be that for the velocity of fractional Brownian motion, since the motion has been derived from the scaled Langevin equation.

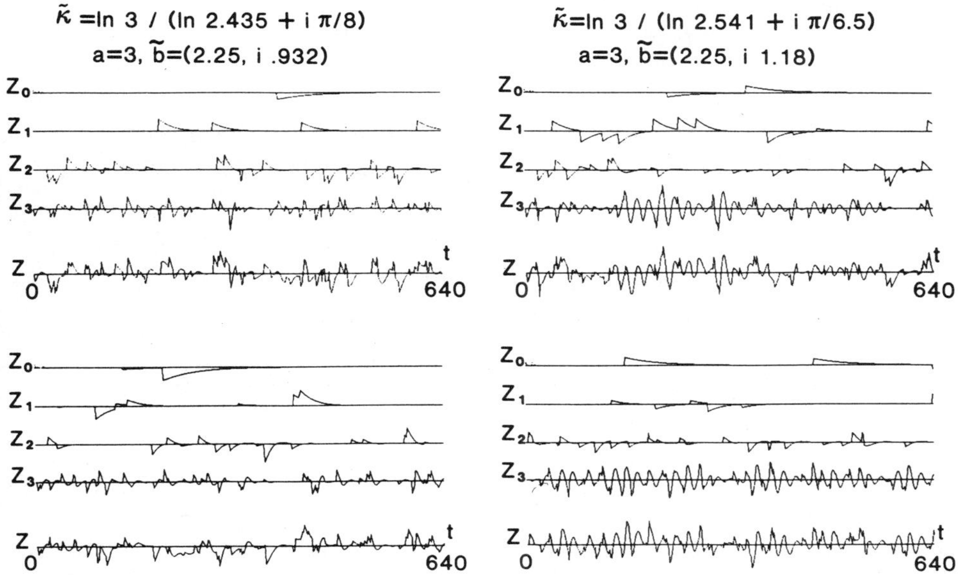

Figure 7 – 6. Complex Brownian motion derived from the Langevin equation by complex-valued scaling. Scaling parameters a and $\tilde{b}$ and complex fractal dimension $\tilde{\kappa}$ are indicated. Two examples for each value of $\tilde{\kappa}$ are simulated. Unit of vertical axis is arbitrary. The first four local states and the response of the whole system are shown. Note that the oscillatory nature of the complex Brownian motion compared with Fig. 7-1.

7.5. COMPLEX SCALING AND RANDOM FAULT PATCHES

The present theory can be extended to describe a more complicated system in which local states are characterized by plural scaling rules. There is no difficulty in the present theory to such multi-scaling problems, provided that the local states are characterized by independent scaling parameters. Here we consider a special case of two scaling parameters, where the basic equations are expressed by complex quantities. A complex-valued scaling constant $\tilde{b}$ is used in a scaling rule for the Langevin equation in this section. This scaling leads to the complex Langevin equation. The solution of a set of complex Langevin equations describes the complex fractional Brownian motion of the Ornstein-Uhlenbeck process as a natural extension of fractional Brownian motion in the previous section.

Complex Langevin equation is important not only because it characterizes the complex Brownian motion but also because it describes a wide variety of physical phenomena. Since the Fourier transform is generally applied to complex functions, functional analysis of the complex Brownian motion is possible. Furthermore, since the complex Langevin equation relates to the motion equation in a general manner, the equation is more fundamental than one in real space.

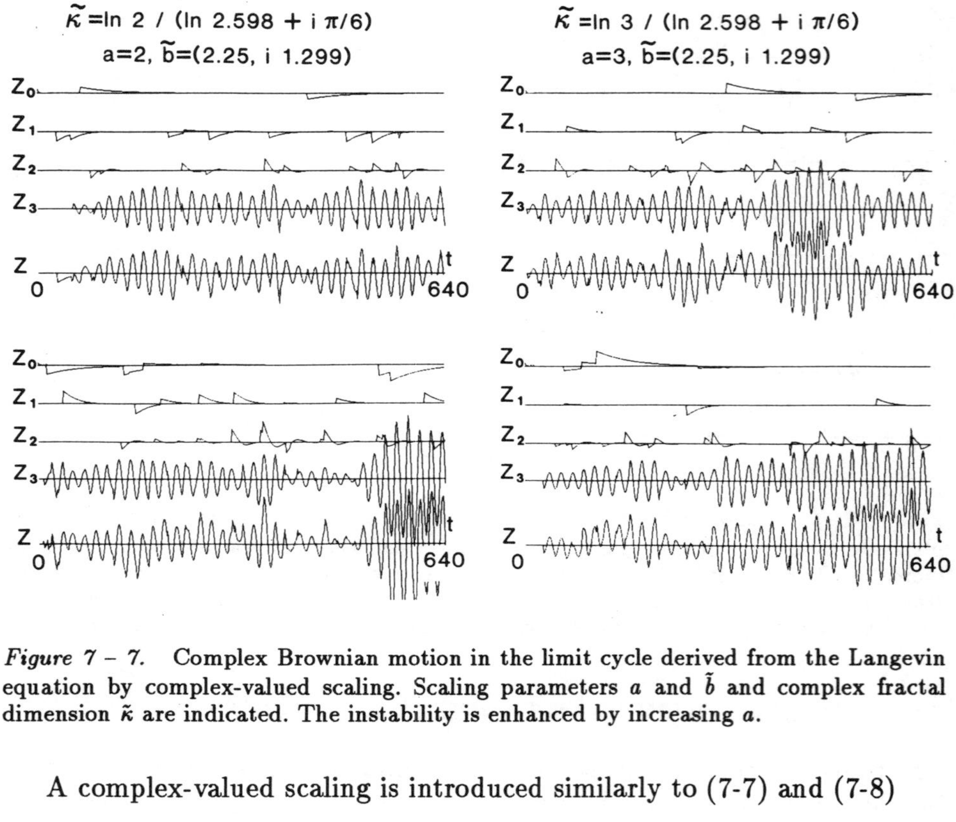

Figure 7 – 7. **Complex Brownian motion in the limit cycle derived from the Langevin equation by complex-valued scaling. Scaling parameters a and $\tilde{b}$ and complex fractal dimension $\tilde{\kappa}$ are indicated. The instability is enhanced by increasing a.**

A complex-valued scaling is introduced similarly to (7-7) and (7-8)

$$f_Z : \sqrt{\frac{a}{\bar{\tilde{b}}}}\; Z_{j-1}(\tilde{b}t) \;\rightarrow\; Z_j(t),$$

$$f_R : \sqrt{a\tilde{b}}\; R_{j-1}(\tilde{b}t) \;\rightarrow\; R_j(t). \qquad (7-33)$$

where $Z_j(t)$ represents the complex-valued j-th local state. The fundamental local state of this system is $Z_0(t) = X_0(t)$. Random force with complex argument is expressed by $R_j(t)$, where $R_0(t) = n_0(t)$.

By the above complex-valued scaling, the Langevin equation is extended to complex space, as

$$\frac{dZ_j(t)}{dt} + \tilde{b}^j \bar{\gamma} Z_j(t) = R_j(t). \qquad (7-34)$$

Solutions of (7-34), $Z_j(t)$ $(j = 0, 1, 2, ...)$, can be shown to be orthogonal. Similarly to (7-10), the scaled random force $R_j(t)$ satisfies an autocovariance function

$$< R_j(t+s)R_j^*(t) >= a < R_{j-1}(t+s)R_{j-1}^*(t) >, \qquad (7-35)$$

where $*$ denotes the complex conjugate. This is a special case of complex Gaussian noise for generating complex Brownian motion. Since the complex Langevin equation by this scaling rule combines the real and imaginary parts, the complex Brownian motion (cBm) shows a coupling between real and imaginary processes. This is the essential difference of the present formalism from previous theories of complex Brownian motion (e.g., Hida, 1980).

Similarly to the real process, the complex-valued system in this section is characterized by a complex fractal dimension $\tilde{\kappa}$ defined by complex scaling parameters as

$$\tilde{\kappa} = \frac{\ln(a)}{\ln(\tilde{b})}. \qquad (7-36)$$

The parameter a relates to an activation rate of events in this complex-valued random system. It determines whether an element is activated or not, so that it must be real. From (7-34), we see that the real part of $\tilde{\kappa}$ measures the scaling length, while the imaginary part describes the scaling frequency: the scaling length corresponds to the pulse width in Fig. 7-1 and the scaling frequency corresponds to the oscillation within the pulse width. Figures 7-6 and 7-7 show simulation examples of the cBm thus obtained. We see the oscillatory behavior of the cBm by the present Langevin formalism. This behavior is enhanced as the scaling proceeds.

The autocovariance of the complex-valued random system can be expressed similarly to the real process, by considering that the initial condition of each local state is $Z_j(0) = 0$. When the real part of the scaling parameter $\Re\{\tilde{b}^j\}$ is positive as the scaling level j changes from 0 to N, it is

$$\begin{aligned} V_{ZZ}(s) &= < \sum_{j=0}^{N} Z_j(t+s) \sum_{k=0}^{N} Z_k^*(t) > \\ &= \frac{\varrho^2}{\bar{\gamma}} \sum_{j=0}^{N} \frac{a^j}{\tilde{b}^j + \tilde{b}^{*j}} \exp(-\tilde{b}^j \bar{\gamma} s) \qquad (s > 0) \\ &\text{or} \\ &= \frac{\varrho^2}{\bar{\gamma}} \sum_{j=0}^{N} \frac{a^j}{\tilde{b}^j + \tilde{b}^{*j}} \exp(\tilde{b}^{*j} \bar{\gamma} s) \qquad (s < 0), \qquad (7-37) \end{aligned}$$

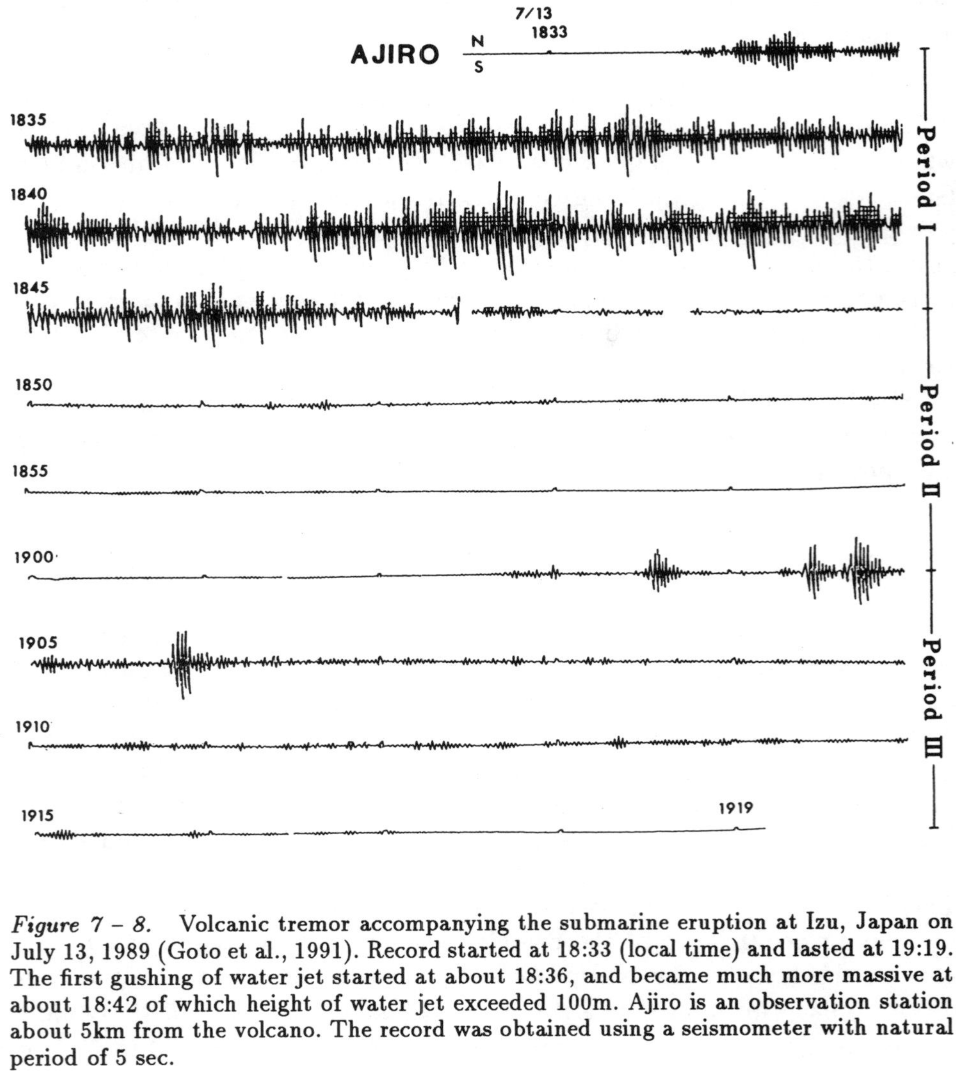

Figure 7 – 8. Volcanic tremor accompanying the submarine eruption at Izu, Japan on July 13, 1989 (Goto et al., 1991). Record started at 18:33 (local time) and lasted at 19:19. The first gushing of water jet started at about 18:36, and became much more massive at about 18:42 of which height of water jet exceeded 100m. Ajiro is an observation station about 5km from the volcano. The record was obtained using a seismometer with natural period of 5 sec.

where $Z_j^*(t)$ is the complex conjugate of $Z_j(t)$ and similarly for $\tilde{b}^*$.

One particular case is $\Re\{\tilde{b}^N\} = 0$ and the others are for $\Re\{\tilde{b}^j\} > 0$ ($j = 0, 1, 2, ..., N-1$). Since all other local states are stationary as in (7-37), we investigate only the behavior of the N-th local state. The autocovariance function of the N-th local state can be evaluated as

$$< Z_N(t_1) Z_N^*(t_2) >= (t_1 \wedge t_2) \exp(-i b_N^I \bar{\gamma} \tau), \qquad (7-38)$$

where the time lag is $\tau = |t_1 - t_2|$ and the imaginary part of the scaling constant is

$$b_N^I = \Im\{\tilde{b}^N\}, \tag{7-39}$$

and the convention $\wedge$ indicates

$$t_1 \wedge t_2 = \begin{cases} t_1 & (t_1 < t_2), \\ t_2 & (t_1 > t_2). \end{cases} \tag{7-40}$$

Figure 7-7 shows the cBm for this case. We see that the complex-valued random system is characterized by a harmonic oscillation with a frequency corresponding to $\bar{\gamma} b_N^I$. Simulations in Fig. 7-7 indicate an instability in Brownian motion, which represents the limit-cycle behavior of the Brownian motion. There is a case $\Re\{\tilde{b}^N\} < 0$ after repeated complex-valued scaling, in which the system loses a linear stability. This suggests the structural instability of Brownian motion.

The complex Langevin equation (7-34) can be rewritten by its real and imaginary parts, $X_j(t)$ and $Y_j(t)$, respectively, as

$$\frac{dX_j(t)}{dt} + b_j^R \bar{\gamma} X_j(t) - b_j^I \bar{\gamma} Y_j(t) = R_j(t), \tag{7-41}$$

$$\frac{dY_j(t)}{dt} + b_j^I \bar{\gamma} X_j(t) + b_j^R \bar{\gamma} Y_j(t) = 0. \tag{7-42}$$

Since equation (7-41) represents an interesting process of the complex system and equation (7-42) shows its background process, the complex-valued scaling combines these two processes and characterizes the coupling between the observable and background processes. Therefore, in this chapter the complex Brownian motion is described not only by the random force and frictional resistance of the classical Brownian motion but also by the feedback effect of the background. A systematic search for such instability in physically relevant situations would constitute an interesting problem. Figure 7-8 shows volcanic tremor accompanying the 1989 Izu submarine eruption, Japan. Monochromatic oscillations with random amplitudes built up over about 10 minutes and the submarine volcano erupted. This would be one example which shows the limit cycle behavior and structural instability of this section.

7.6. SPECTRUM OF COMPLEX BROWNIAN MOTION

Since the autocovariance in (7-37) is $V_{zz}(s) = V_{zz}^*(-s)$, the power spectrum is real:

$$P_{cBm}(\omega) = \frac{\varrho^2}{\bar{\gamma}} \sum_{j=0}^{N} \frac{a^j}{\tilde{b}^j + \tilde{b}^{*j}} \left\{ \frac{1}{\bar{\gamma}\tilde{b}^j + i\omega} + \frac{1}{\bar{\gamma}\tilde{b}^{*j} - i\omega} \right\}. \tag{7-43}$$

A condition of $\Re\{\tilde{b}^j\} > 0$ $(j = 0, 1, 2, ..., N)$ is considered in this case. In order to study asymptotic behavior of the above spectrum, a scaling relation is considered similarly to (7-21). This is to replace ω by $\frac{\omega}{\tilde{b}}$ and $\frac{\omega}{\tilde{b}^*}$ in the right hand side of (7-43). Provided that $|\Im\{\tilde{b}\}| < \tilde{b}\tilde{b}^*$ and $a < \tilde{b}\tilde{b}^*$, the spectrum is approximated as

$$P^S_{cBm}(\omega) \simeq \tilde{A}_{\tilde{\kappa}}\, \omega^{\tilde{\kappa} - 1 - \ln \tilde{b}^* / \ln \tilde{b}} + \text{c.c.}, \qquad (7-44)$$

where $\tilde{A}_{\tilde{\kappa}}$ is a constant depending on the complex fractal dimension $\tilde{\kappa}$, and c.c. indicates the corresponding complex conjugate. The power spectrum (7-44) is expressed by a power-law of frequency, and the exponent coefficient is a function of the complex fractal dimension. Considering that the imaginary part of $\tilde{b}$, $\Im\{\tilde{b}\}$ tends to 0, the spectrum (7-44) is equivalent to (7-19) as expected.

When the scaling parameters satisfy a condition of $a > \tilde{b}\tilde{b}^*$, an approximation of (7-43) is evaluated as

$$P^A_{cBm}(\omega) \simeq \varrho^2 \frac{a^N}{(\bar{\gamma}\tilde{b}^N + i\omega)(\bar{\gamma}\tilde{b}^{*N} - i\omega)}, \qquad (7-45)$$

where $|\tilde{b}^* - \tilde{b}| \ll \tilde{b}\tilde{b}^*$ has been assumed. In the limit of $\Im\{\tilde{b}\} \to 0$, this spectrum is equivalent to the Lorentz spectrum for real processes.

The limit cycle behavior of the complex-valued random system also occurs. The autocovariance (7-38) in the limit cycle is apparently not a function of the time lag $t_1 - t_2$. Therefore the system is not stationary. A spectral theory for nonstationary stochastic processes provides an estimation of an average power spectrum: Suppose that a random force lasting from $t = 0$ to $t = T$ is considered as input to a particular system. The corresponding output lasts from 0 to T. The average energy spectrum of the output is expressed by the product of the average energy spectrum of the input random force and the squared Fourier transform of the impulse response of the system. Since a power spectrum is generally defined by an energy spectrum per unit time, the power spectrum of the complex Brownian motion in the limit cycle is expected to be

$$P^L_{cBm}(\omega) \simeq \varrho^2 a^N \{\delta(\omega + \bar{\gamma}b^I_N) + \delta(\omega - \bar{\gamma}b^I_N)\}, \qquad (7-46)$$

where δ is Dirac delta function. This indicates a spectrum of harmonic oscillations with a characteristic frequency of $\bar{\gamma}b^I_N$.

7.7. RANDOM FAULT PATCHES AND FRACTIONAL BROWNIAN MOTION

Real and complex-valued scaling parameters have been considered to define the fractal dimension and the complex fractal dimension of fractional

Brownian motion. Since each scaling characterizes a fine structure of local states, the present scaling corresponds to the fine-graining operation (opposite to the coarse graining) of Brownian motion as the computer simulations indicate.

The complex-valued scaling for the Langevin equation demonstrates fractional Brownian motion coupled with the relevant background. The scaling procedure will result in Brownian motion in the limit cycle, where the periodicity of the response is sensitive to the coupling between real and background processes. Furthermore, the instability is extended to the qualitative change in Brownian motion by complex-valued scaling. This is a significant result of complex-valued scaling applied to the Langevin equation and extends previous representations of the complex Brownian motion.

Computer simulations show that the real and complex-valued scalings describe the system responses by random-pulses and random oscillatory-pulses. These random-pulses and random oscillatory-pulses compose an orthonormal system of functions. These responses generally describe the dynamics of seismic pulses radiated from the rupture process of random fault patches. Thus the present theory is a candidate for representing practical phenomena of complex systems with fine-structures, since the orthonormal functions can represent the stochastic process of random activations in a general manner.

The essential result from this chapter is the three modes of steady-state random systems characterized by $1/f$ spectrum, Lorentz spectrum and Kolmogorov spectrum. These spectra have been shown to be a steady state rupture process of random fault patches, random and dense ruptures, and random but sporadic ruptures, respectively.

CHAPTER 8

STOCHASTIC MODELING OF COMPLEX EARTHQUAKE ACTIVITY

The random occurrence of earthquakes is investigated in this chapter in relation to the stochastic scaling of earthquake activity. In the first approach we introduce a mathematical model that describes temporal variations of earthquake activity. This model regards earthquake activity as a point process composed of earthquake events without any size effect. Effects of hysteresis and time delays on earthquake occurrences are investigated. This model reproduces aftershock series, earthquake swarms, and periodic patterns of earthquakes. When an external periodic effect is considered, the model is further generalized. Various non-periodic as well as chaotic patterns of complex earthquake activity can be simulated.

In the second approach we consider a cluster of many series of earthquake occurrences classified by different magnitudes. Each series of earthquakes can be regarded as a point process with a particular source-size effect, because of their common magnitude. Although each series of earthquakes occurs more or less at random, the stochastic properties are not the same among different series of earthquake occurrences. This is mainly due to the stochastic size-effect of the complex earthquake activity. The last effect is considered in the fundamental equation of earthquake activity by extending the concept of the stochastic scaling.

In this chapter stochastic scaling is further investigated in order to clarify the universal nature of earthquake activity as a complex system. This stochastic scaling is not a scaling relation in fractal geometry or the scale invariant nature but specifies the scaling of statistical moments among different random processes classified by different energy levels of events.

8.1. PERIODIC-APERIODIC EARTHQUAKE ACTIVITY

The earthquake source is represented by a complex faulting process composed of random ruptures of small-scale fault heterogeneities. The fluctuation of heterogeneous stresses on a fault plane is very important for the generation of an *isolated* earthquake, mainshock-aftershocks and an earthquake swarm (Cheng and Knopoff, 1987). The spatial distribution of these fault heterogeneities represents locked and unlocked fault patches (fault asperities and barriers) on the heterogeneous fault plane. These fault patches appear to exist on all scales producing stress concentration locally on the fault.

The stress concentration around fault patches becomes the source area of aftershocks. Aftershocks following large shallow earthquakes are considered to be the relaxation of such stress concentrations induced by a particular mainshock. The time evolution of the relaxation process is described empirically by Omori-Utsu's empirical formula as a hyperbolic decay rate with time

$$n(t) = \frac{A}{(t+B)^p}, \tag{8-1}$$

where n is the number of aftershocks at a time t after the origin time of

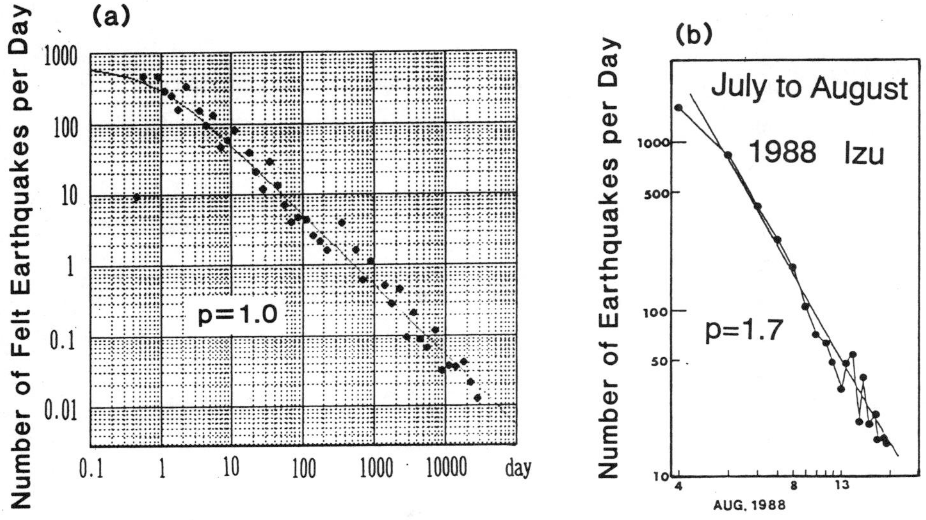

Figure 8 – 1. **Number of earthquakes per day represented by a power law. (a) The number of felt earthquakes after the 1891 Nobi earthquake, Japan (Utsu et al., 1995) and (b) the 1988 earthquake swarm near Izu peninsula, Japan (Tokyo University, 1989). Power coefficient p of the decay rate is indicated.**

the mainshock, and A and B are positive constants. The power coefficient p in (8-1) has been determined to be about 1.0 to 1.5 for large earthquakes in the world. An example is shown in Fig. 8-1(a) counting the number of felt earthquakes after the Nobi earthquake of 1891 in Japan. p in this case is about 1.05 and not rigorously one for the aftershock activity but such value for p has been measured more than 100 years. Also p for earthquake swarms is usually larger than 1. An example is shown in Fig. 8-1(b) for the 1988 earthquake swarm near Izu peninsula, Japan, for which p is as large as 1.7. Omori-Utsu's empirical formula is different from the simple exponential decay of events as a function of time by Poisson process, i.e., the formula is representing one aspect of the complexity of earthquake phenomena.

In the case of earthquake swarms the number of earthquakes increases initially and then decreases gradually with time. An exponential time-decay was seen for the Matsushiro earthquake-swarm activity in Japan. Exponential time-decay has also been found for earthquake swarms in East-off Izu peninsula, Japan. Figure 8-2 shows this behavior. An exponential time-decay indicates that such activity is a Poisson process. We do not know any physical relationship between the exponential time-decay of earthquake swarms and the power-law decay of aftershocks, although some studies derive such a relation based on physically-unproved a priori assumption.

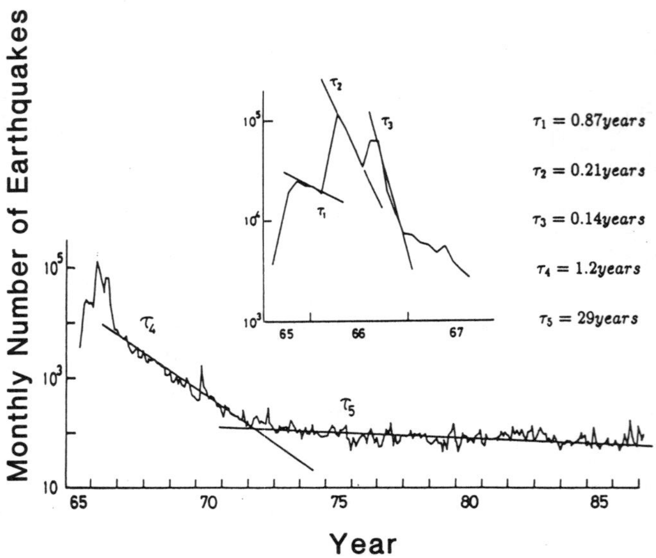

Figure 8 – 2. Monthly number of earthquake occurrences in Matsushiro, Japan represented by an exponential time-decay (Tsukuda, 1993).

For external periodic force which might activate earthquakes, the largest is the tidal force. The correlation between the earth tides and earthquake activity is an intriguing subject in seismology. The most pronounced correlation has been found for deep-focus moonquakes observed by the Apollo passive seismic experiment on the moon. Figure 8-3 shows occurrence times of one family of moonquakes which occurs at the same source area (Nakamura, 1978) plotted on the trace of locus of the sub-earth point‡. The subearth point moves due to the monthly libration of the moon. Periodicity of moonquake occurrences as well as the time-dependent force system to generate the activity have been pointed out (Koyama and Nakamura, 1980).

A prominent diurnal periodicity has been found in earthquake swarms in Kilauea volcano, Hawaii, however the semidiurnal periodicity is not clear. This is surprising because the semidiurnal earth-tide dominates in Hawaii.

‡Sub-earth point is the point where the imaginary line connecting the mass centers of the earth and the moon intersects the lunar surface.

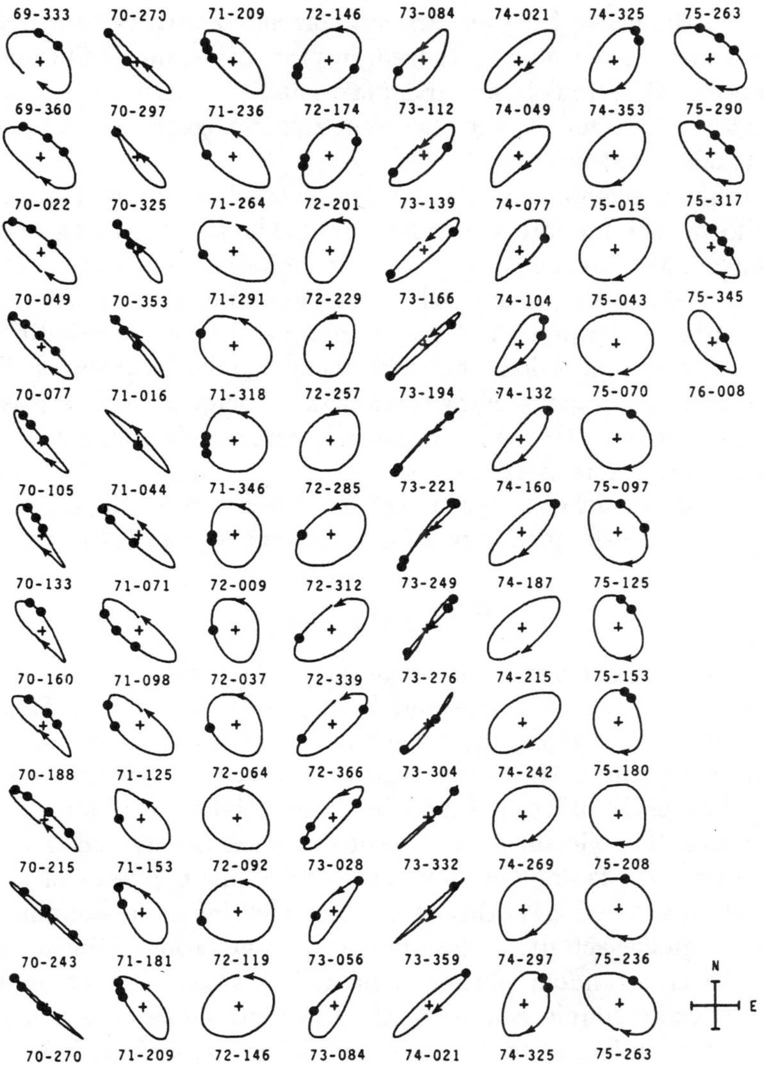

Figure 8 – 3. Occurrence times, plotted relative to the libration of the moon, of A_1 moonquakes (a family of deep-focus moonquakes) observed by the Apollo passive seismic experiment. The location of the sub-earth point at the time of occurrence of each A_1 moonquake is indicated by a solid circle. The sub-earth point is the locus of intersections of the imaginary straight line connecting the mass centers of the earth and the moon with the lunar surface. The sub-earth point moves on the lunar surface each month as a result of the latitudinal and longitudinal librations of the moon. Each figure covers an anomalistic month from apogee to apogee, with an arrow placed at the end of the month. The numbers between the figures are the year and the Julian day of apogee crossing. Cross at the bottom right corner indicates the scale of ±5 degrees (Nakamura, 1978).

The moonquake activity evidently shows a close correlation with the earth-moon tide not only in occurrence but also in terms of their force systems. Natural earthquakes, however, are not correlated with the earth tides nor do they show the periodicity. Indeed, in most cases, no significant correlation between disastrous large earthquakes and the earth tide can be found. Although the tidal forces may play some role in triggering earthquakes, the effect remains controversial.

Earthquake phenomena are very complicated and many factors such as non-uniform tectonic forces, temperature fields, and the earth's structures are involved. Mechanisms to generate the periodicity of earthquake activity will be considered mathematically in the following sections. We will show that modulation by periodic forces sometimes produces periodic patterns and other times non-periodic patterns of earthquake occurrences. Apparent varieties and irregularities of the earthquake activity are due, at least partly, to the competition between the periodic force such as the tide and the autonomous rhythms of earthquakes.

The Goishi model of Ohtsuka (1972) is purely stochastic, and provides a similar magnitude-frequency relation to Gutenberg and Richter's empirical relation (1954) of

$$\log N(M_S) = a - bM_S, \qquad (8-2)$$

where N is the number of earthquakes larger than surface-wave magnitude M_S in a certain region, and a and b are positive constants. This empirical relation shows that the number-size distribution of natural earthquakes is a power law since M_S is the logarithm of earthquake source strength. This is apparently different from the exponential form of simple stochastic processes. The Goishi model is not based on the dynamics of rupture propagation on a fault plane but on a probabilistic growth of a tree-like shape. Self-organized criticality is a recent idea for understanding complicated earthquake activity. These numerical simulations indicate that the earthquake is a random phenomenon with a small number of freedoms under a specific simple rule and that the earthquake as a complex system is in the critical state of phase transitions. Therefore, it is tempting to search for a universality in the fundamental physical origin of complex earthquake activity in relation to the complex system in physics. In this chapter the complexity of earthquake activity is investigated theoretically and physically in more detail to elucidate the fundamental rule underlying the complex earthquake activity.

8.2. EARTHQUAKE ACTIVITY AS A POINT PROCESS

Here we consider time variation of earthquake occurrences in a given region theoretically and dynamical modeling is proposed for the earthquake phenomena. The number of earthquakes is the preferred parameter for describing the earthquake activity since it is the most reliable and easiest available data. Moreover, several empirical formulae are known for the temporal variation representing the complexity of the earthquake activity.

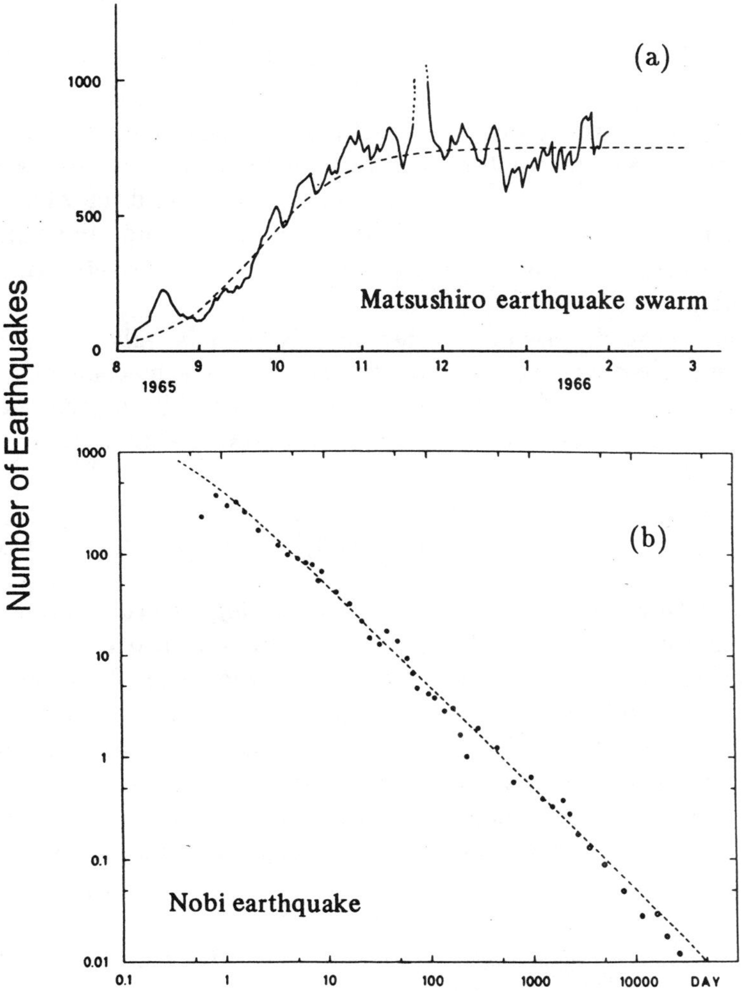

Figure 8 – 4. Earthquake activity described by a logistic equation. (a) The Matsushiro earthquake swarm from 1965, and (b) aftershock activity of the 1891 Nobi earthquake in Japan. Dashed curves in the figure denote the theoretical logistic growth of the corresponding activity (Ouchi, 1982) using different initial conditions.

We consider a certain region as a system which generates earthquakes. This system is, in general, highly complicated and effects of various circumstances must be considered. Because of external effects, this is essentially an open system. Since it seems almost impossible to formulate the dynamics of such a complicated system completely, we start from the deterministic viewpoint and ignore the various external effects for the sake of simplicity. We assume the following rather general dynamical equation to describe the temporal variation of earthquake occurrences in this region:

$$\frac{dn(t)}{dt} = F(n, c_1, c_2, c_3, \ldots), \qquad (8-3)$$

where $n(t)$ denotes the number of earthquakes at a time t and $c_i (i = 1, 2, 3, \ldots)$ are various parameters such as heterogeneous stresses, temperatures, and structures which characterize the function F and reflect the state of the system. Since function F is highly complicated and equation (8-3) must be non-linear, complete formulation would be excessively difficult, if not impossible.

To advance the discussion, we assume F depends only on $n(t)$. The function is regarded as positive and negative feedback mechanisms of the system. We consider further a function F in which the system behaves *chaotically* in the next section. For now F is represented by a Taylor series of n as

$$F(n) = n(\alpha_0 - \beta_0 n), \qquad (8-4)$$

where α_0 and β_0 are positive constants, and the higher order terms of the Taylor series have been neglected (Ouchi, 1982). Physically, α_0 describes the production rate of earthquakes and β_0 the reduction. Equation (8-4) leads (8-3) to the logistic equation which has been used to describe the population dynamics of bio-species. In the logistic equation, all earthquakes are numbered irrespective of their magnitudes or of their source sizes. Therefore, the system in this section is essentially a point process. Equation (8-3) under the condition of (8-4) can be solved analytically and the solution is expressed as

$$n(t) = \frac{\alpha_0}{\beta_0} \frac{n_0}{(\frac{\alpha_0}{\beta_0} - n_0)\exp(-\alpha_0 t) + n_0} \quad (t \geq 0). \qquad (8-5)$$

This solution shows two types of time variations depending on the initial value of n_0:

(a) If $n_0 < \frac{\alpha_0}{\beta_0}$, the number of earthquakes gradually increases with time and then approaches a constant value. An example of this solution is illustrated in Fig. 8-4(a) compared with observations of the Matsushiro earthquake swarm in Japan.

(b) If $n_0 > \frac{\alpha_0}{\beta_0}$, the number of earthquakes shows a power-law like decrease of activity as in Fig. 8-4(b) for the aftershock series of the 1891 Nobi earthquake in Japan. This temporal variation is quite similar to that in Fig. 8-1 described by (8-1). In Fig. 8-4(b) the decrease of earthquake occurrences over 100 years can be explained by the solution of (8-3) as well as by the empirical power law of Omori and Utsu. The solution (a) describes the temporal variation of particular earthquake swarms and the solution (b) that of aftershock sequences in a general manner, even though the basic process is assumed to be a point process without any size effect.

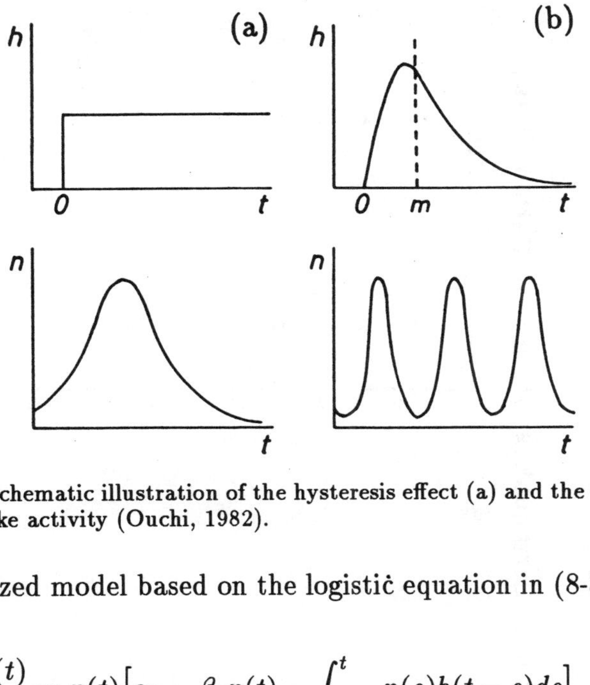

Figure 8 – 5. Schematic illustration of the hysteresis effect (a) and the time delay effect (b) on earthquake activity (Ouchi, 1982).

A generalized model based on the logistic equation in (8-3) is proposed as

$$\frac{dn(t)}{dt} = n(t)\Big[\alpha_0 - \beta_0 n(t) - \int_{-\infty}^{t} n(s)h(t-s)ds\Big], \qquad (8-6)$$

where $h(t)$ denotes the hysteresis effect of earthquake occurrences. Various hysteresis effect on earthquake activity can be represented by $h(t)$. For example, if $h(t)$ =constant and β_0=0 ($t > 0$), then the influence accumulates with time due to the hysteresis and the number of earthquakes gradually increases and then decreases (Fig. 8-5(a)).

A more realistic hysteresis is the case $h(t) = \kappa_0\delta(t-m)$ and $\beta_0 = 0$, in which the time delay effect is modeled by Dirac delta function $\delta(t-m)$ with an amplitude of κ_0. The delayed logistic equation is then expressed as

$$\frac{dn(t)}{dt} = n(t)[\alpha_0 - \kappa_0 n(t-m)]. \qquad (8-7)$$

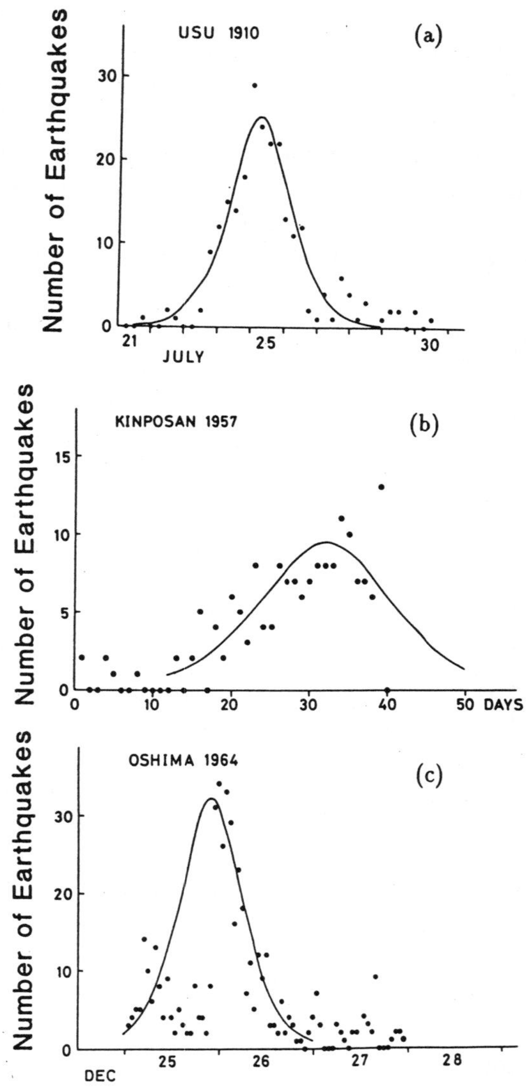

Figure 8 – 6. Temporal variation of earthquake swarm activity around (a) Usu volcano in 1910, (b) Kinposan in 1954, and (c) Oshima in 1964. The solid curves denote theoretical activity calculated from (8-6) for each case (Ouchi, 1993).

This equation has been originally considered for describing the population dynamics of bio-species. The solution of (8-7) shows periodic oscillation or limit cycle behavior.

A more gradual time-delay has been considered for the genetic dynamics of bio-species by introducing a smooth function for the time delay effect as

$$h(t) = \kappa_0[\exp(-\gamma_1 t) - \exp(-\gamma_2 t)] \quad (\gamma_2 > \gamma_1). \qquad (8-8)$$

Two exponential terms in (8-8) can be interpreted as dissipation and activation of the earthquake activity as mentioned before, where γ_1 and γ_2 represent different rates for these effects. Numerical examples of the hysteresis effect for (8-8) can be found in Fig. 8-5(b). The solution shows a rhythmic (periodic) occurrence of earthquake bursts.

Figure 8-6 shows the temporal variation of observed earthquake sequences including those of volcanic origins and the corresponding simulation by (8-6). Some earthquake swarms in the vicinity of volcanoes show remarkable periodicities, which are also found in microearthquakes. This kind of rhythmic earthquake occurrence may be a rather general phenomenon, particularly for volcanic earthquakes. A physical mechanism for the time delay effect in (8-7) to produce such autonomous rhythms would be a coupling of the earthquake activity to its background environment. For instance, coupling among tectonics stresses, visco-elastic stresses, tidal stresses, and magmatic forces would activate earthquakes.

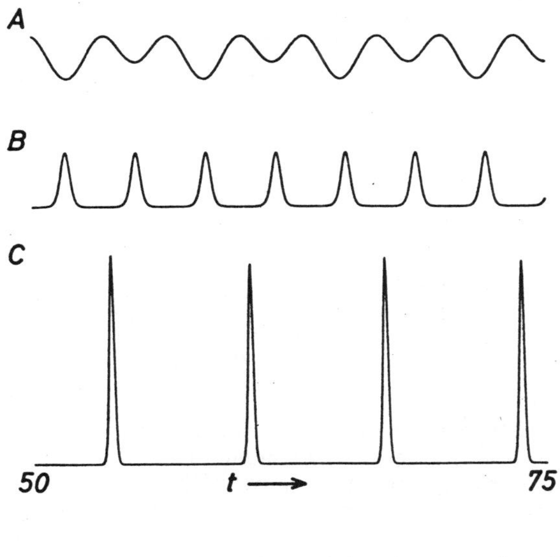

Figure 8 – 7. External effect on earthquake activity. A: external periodic force $F(n,t)$; B: original rhythmic pattern of earthquake activity; and C: resultant earthquake activity $n(t)$. Note that bursts of earthquakes occur exclusively at a certain phase of the external periodic force (Ouchi, 1993).

8.3. EXTERNAL EFFECT ON EARTHQUAKE ACTIVITY

The external random force is also considered in the logistic equation of (8-3). Let external force $F(n,t)$, equation (8-3) be rewritten as

$$\frac{dn(t)}{dt} = n(t)\Big[\alpha_0 - \beta_0 n(t) - \int_{-\infty}^{t} n(s)h(t-s)ds\Big] + F(n,t). \quad (8-9)$$

When $F(n,t)$ is random force, the solution of (8-9) exhibits a repetitive pattern activated at random as discussed in the previous section. If we consider a periodic external force, we can assume $F(n,t)$ to be

$$F(n,t) = n\{r_1 \sin(2\pi f_1 t + \theta_1) + r_2 \sin(2\pi f_2 t + \theta_2)\}. \quad (8-10)$$

This is an approximation of the tidal effect. Two periods in (8-10) are taken to represent the diurnal and semidiurnal components of the tide, for example. Here, f_1, f_2, θ_1, θ_2, r_1, and r_2 are frequencies, phases and amplitudes of these two periodic components, respectively.

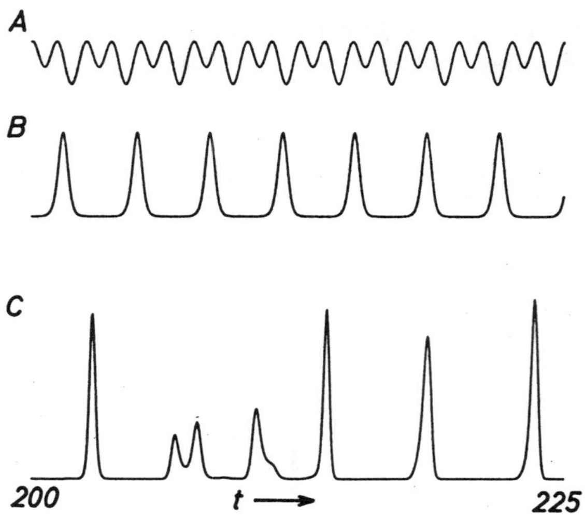

Figure 8 - 8. External effect on earthquake activity. Symbols are the same as those in Fig. 8-7 except for constant parameters of amplitudes and periods of external force and original rhythmic pattern. The resultant activity C is shown to be *chaotic*, because the Lyapunov number of this is positive with a value of 0.233 (Ouchi, 1993).

Numerical solutions of (8-9) provide various interesting patterns. In some cases, phenomena are *synchronized* by the periodic external force (Fig. 8-7). In this case, earthquake bursts occur exclusively at a certain phase of the external force and the activity has a period synchronized with the longer period of assumed frequency components in (8-10). This explains the observation of diurnal tidal periodicity in earthquake occurrences and no observation of the semidiurnal tidal periodicity for volcanic earthquakes.

Chaotic and non-periodic behavior appears under certain conditions in the numerical solution of (8-9), even though a periodic external force is applied. Figure 8-8 gives an example of numerical simulations for such cases. This result describes the non-periodic and/or random occurrences of earthquakes due to periodic generating sources. This kind of phenomena is quite common in nonlinear systems in physics and is dependent both on the period of the original sequence of earthquakes and on the amplitude and period of external forces in (8-10).

8.4. STOCHASTIC SCALING OF EARTHQUAKE ACTIVITY

There occur a number of aftershocks following each large earthquake. Aftershocks distribute on a fault plane of the mainshock more or less in a random manner. The time sequence of aftershock occurrences is not periodic but sporadic. Studies of the statistical nature of earthquakes have tested for a Poisson process and for a Markov process. These studies view the earthquake source as a point without any characteristic energy nor finite source size.

The number density of aftershocks decreases systematically as a power law described by (8-1) and does not decrease in a statistically homogeneous manner. Furthermore, the size of earthquakes varies by many orders of magnitudes. Some are large enough to be secondary disastrous earthquakes and some are so small that can be detected only by sensitive instruments. The randomness of occurrence times and the size distribution are inherent to the earthquake source properties, and must be considered in order to obtain a complete understanding of the earthquake statistics. This viewpoint is quite different from the one for earthquake swarm activity in the previous section and from previous studies.

The number $n(t)$ of aftershocks in (8-1) is the sum of aftershocks with different earthquake-magnitude ranges as

$$n(t) = \sum_{i=0}^{M} n_i(t), \qquad (8-11)$$

where $n_i(t)$ represents one series of aftershocks classified by an i-th earthquake-magnitude range at a time t. The earthquake-magnitude range specified by $i = 0$ is the series with the minimum magnitude observed in a particular earthquake activity and $i = M$ indicates that with the maximum magnitude. Because of the randomness, $n_i(t)$ and $n_j(t)$ are independent, provided that $i \neq j$. This type of size-dependent randomness has been neglected, although it is very important to account for the scaling and the energy of the earthquake activity. Suppose that there are left $N_i{}^0$ nuclei produced by the mainshock for succeeding aftershocks of the i-th magnitude range. The number of aftershocks of the i-th magnitude range within the time interval $t \sim t + dt$, $n_i(t)dt$, can be written as

$$n_i(t)dt = -dN_i(t) = \mu_i(t)N_i(t)dt, \qquad (8-12)$$

where $N_i(t)$ is the number of nuclei classified by the i-th magnitude range to be ruptured after time t and $\mu_i(t)dt$ is the probability of rupturing one nucleus in that time interval. The initial condition for (8-12) has been given by $N_i{}^0$ at $t = 0$. In general, there may be a weak time-dependency and also magnitude dependence of $\mu_i(t)$. Some studies a priori assume such dependence without any physical justification. However, we assume $\mu_i(t)$ to be a constant of μ_0 throughout this study for the sake of simplicity.

We may not be able to characterize the stochastic size-effect of earthquake activity by such a small number of macroscopic parameters. We consider that each series of the earthquake activity is governed by the same equation of the temporal evolution in (8-12) with respective stochastic behavior. A mathematically simple assumption is to derive a respective scaling relation for N_i's, though they are random functions and statistically independent. A stochastic scaling has been introduced to specify the scaling relation for such random functions in the previous chapter

$$f_d : \frac{a_d}{b_d} N_i(b_d t) \rightarrow N_{i+1}(t), \qquad (8-13)$$

where a_d and b_d are scaling parameters, positive constants, and both are smaller than unity. When $a_d = b_d$, the stochastic scaling is self-similar. When $a_d \neq b_d$, it is self-affine. However, this scaling of f_d does not specify the self-similarity and/or the self-affinity of functional forms, but represents a mapping which describes the similarity nature underlying the statistical properties of N_i's. Scaling in a fractal geometry and in a scale invariant nature introduces a relationship of some function $f(x)$ as $f(ax) = a^H f(x)$, where H and a are positive constants. The stochastic scaling in this study specifies a scaling relation in the fundamental equation of earthquake activity, such as (8-12). Therefore, this scaling is designated as *stochastic.*

The solution of (8-11) and (8-12) is now straightforward under the stochastic scaling of (8-13) and it describes the number of aftershocks at a time t after the mainshock as

$$n(t) = -\frac{d}{dt}\Big[\sum_{i=0}^{M} N_0{}^0 \left(\frac{a_d}{b_d}\right)^i \exp(-b_d{}^i \mu_0 t)\Big]. \qquad (8-14)$$

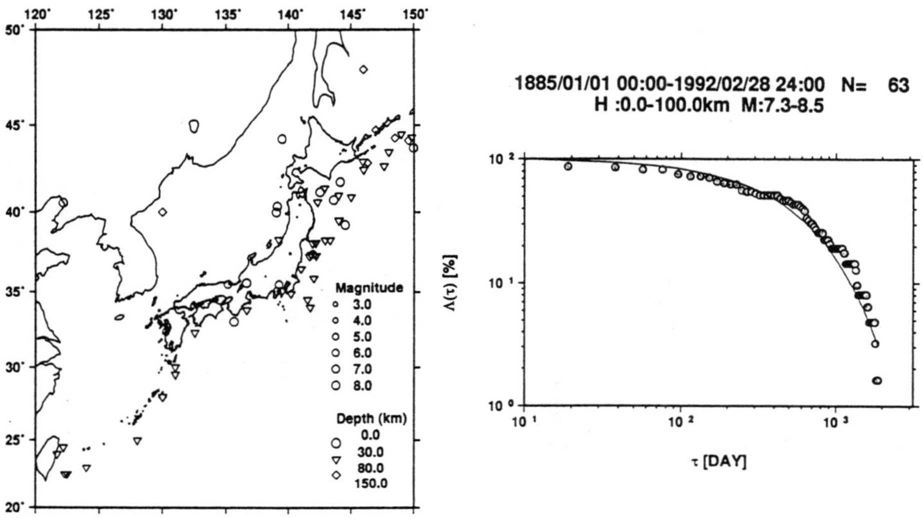

Figure 8 – 9. Time intervals of successive shallow earthquakes in Japan. Large and great earthquakes from 1885 to 1992 are selected with magnitudes larger than or equal to 7.3. Epicenters are plotted by different symbols representing focal depths.

When $0 \leq \mu_0 t \ll 1$, equation (8-14) is related to the assumed initial condition. When $\mu_0 t \gg 1$, an asymptotic form of the solution can be obtained by the aid of the steepest descent method in Appendix G. Then the solution of (8-14) is rewritten

$$n(t) \simeq A_d(\mu_0\, t)^{-\xi_d}, \tag{8 – 15}$$

$$A_d = \mu_0 {N_0}^0 \sqrt{\frac{\pi}{2\ln(\frac{a_d}{b_d})\ln(b_d)}}\,(\xi_d - 1)\exp([\xi_d - 1][\ln(\xi_d - 1) - 1]), \tag{8 – 16}$$

where ξ_d is a fractal dimension defined by the scaling parameters as

$$\xi_d = \frac{\ln(a_d)}{\ln(b_d)}. \tag{8 – 17}$$

The asymptotic form (8-15) is also valid for both a_d and b_d larger than unity. The solution (8-14) indicates the exponential decay of an earthquake activity, whereas its asymptotic form (8-15) manifests the power-law decay. These solutions surely give an insight into the Omori-Utsu's empirical decay of aftershocks and the exponential decay of earthquake swarms in Figs. 8-1 and 8-2. A relationship between the Omori-Utsu's empirical constant p and the fractal dimension of the stochastic scaling is derived as

$$p = \xi_d. \qquad (8-18)$$

This relationship shows the physical meaning of the empirical power coefficient p in terms of the scaling parameters of complex earthquake activity.

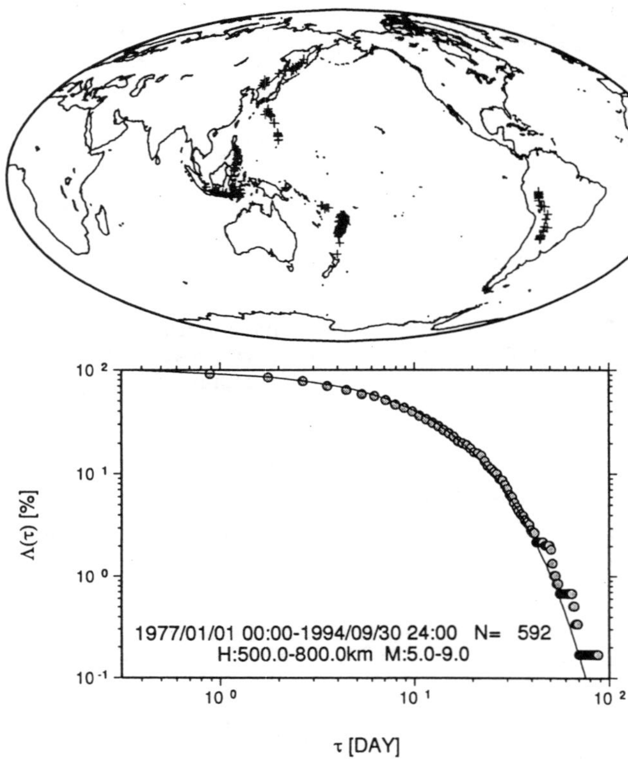

Figure 8 – 10. Time intervals of successive deep-focus earthquakes worldwide. Magnitudes are larger than 5.5 and focal depths more than 500km. Earthquakes are from 1977 to 1994 as tabulated by the U.S. Geological Survey.

8.5. LONG-TAIL BEHAVIOR OF EARTHQUAKE ACTIVITY

The statistical behavior of earthquakes has been investigated for many years. For many cases of large earthquake activities a Poisson distribution fits the observed frequency. In this case, the earthquakes are mutually independent. Since such behavior is found for the data of about 100 years of modern seismometry, the probability of earthquake occurrence is considered to be independent of time and of total number of events. Figure 8-9 shows the distribution of time intervals of successive large shallow-earthquakes in and near Japan. Figure 8-10 shows the same plot for large deep-focus earthquakes worldwide. Poisson distributions, drawn by solid curves in the figures, explain the observed data, demonstrating the randomness of large earthquakes.

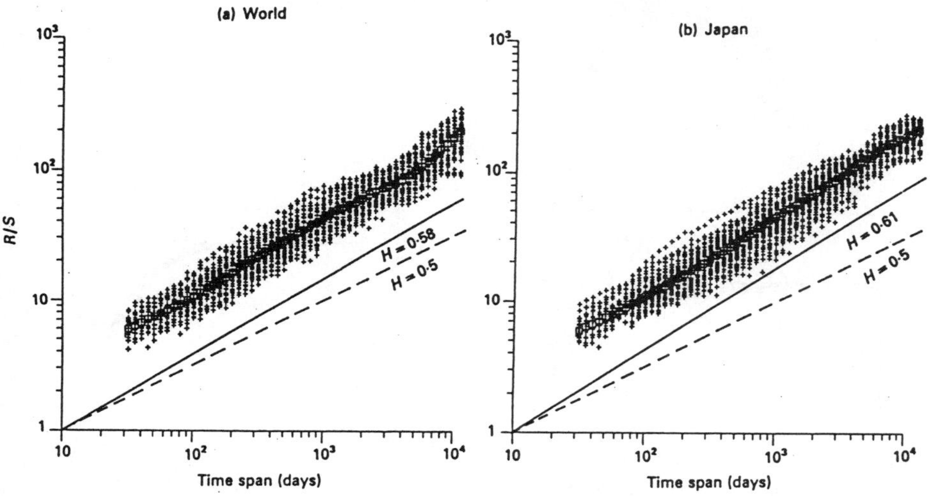

Figure 8 - 11. Autocovariance of global seismicity and for Japan (Ogata and Abe, 1991). H is an empirical constant which characterizes the autocovariance function $|\tau|^{2H-2}$. The same power law holds for seismicity in and near Japan as for worldwide, although the magnitude and number of earthquakes analyzed are very different.

This is consistent with a physical view point. Since the earthquake source is within a large energy reservoir under tectonic loading, the original condition is not much affected by the occurrence of an earthquake and is rapidly recovered, since the energy dissipated by an earthquake is very small. We saw this in discussing the seismic efficiency of the frictional faulting process. Therefore, the earthquake occurrence does not affect the probability of subsequent earthquakes. However, the statistical analysis may be biased if aftershocks and/or earthquakes with small magnitude are included. In many studies aftershocks are removed from earthquake catalogs subjectively or on the basis of experience in order to study the statistics of earthquake occurrence.

Evidence for the long-tail behavior of earthquake activity has been presented by Ogata and Abe (1991) by investigating the autocovariance of

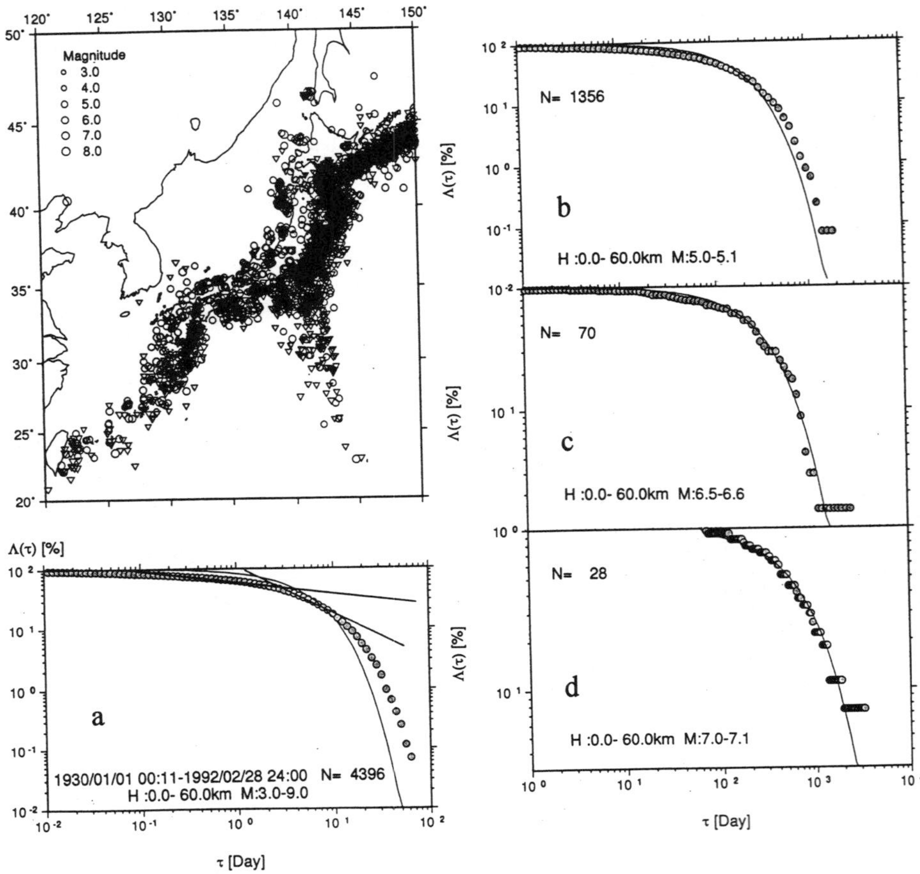

Figure 8 – 12. Cumulative number of time intervals of successive shallow earthquakes in Japan and vicinity. Number of earthquakes N is normalized to represent the probability distribution in %. The earthquakes are from the Japan Meteorological Agency catalog for the period from January 1st, 1930 to February 28th, 1992 and have focal depths less than or equal to 60 km. (a) All earthquakes in the analysis. The curve in the figure is the Poisson probability distribution calculated from the mean number of earthquakes. It does not fit the observed time-interval distribution. Two straight lines are for a power-law dependence between time interval and cumulative number in days. (b) Similar plot for earthquakes in the magnitude range from 5.0 to 5.1. (c) Earthquakes in the magnitude range from 6.5 to 6.6. (d) Earthquakes in the magnitude range from 7.0 to 7.1.

world-wide seismicity (Fig. 8-11). They showed that the autocovariance is represented by a power law $|\tau|^{2H-2}$, where τ is the time lag and H is an empirical constant of about 0.6 (theoretically $0 < H \leq 1$). They also found the same power law for seismicity in and near Japan, where again H is about 0.6 (Fig. 8-11), although the two data sets differ in magnitude and number of earthquakes.

We investigate the time-interval distribution of successive events in and near Japan. Figure 8-12 shows an example of the probability estimated from the cumulative number of time-intervals of successive earthquake events. Because of historical changes in seismic detectability, the minimum magnitude of 5.0 (Japan Meteorological Agency scale) and focal depths shallower than or equal to 60 km are used to evaluate the time-interval distribution. Although these restrictions are subjective, the result does not change very much with change in the numerical values for these restrictions. The result is that the distribution for a particular earthquake activity classified by a narrow magnitude range can be described by a Poisson distribution. A distribution similar to the Poisson can be found not only for the small magnitude range of about 5 but also for the large magnitude range of about 8. This agrees with the result in Fig. 8-9 for the largest earthquakes. However, the time-interval distribution of all the magnitude ranges of earthquakes is no longer consistent with a Poisson distribution. The decay rates of small and medium time-intervals are very small compared with the Poisson. Several percent of the distribution in the large time-interval may be explained by a Poisson distribution. We find a Poisson-like distribution for each earthquake activity classified by each narrow magnitude range, in which only the mean value of earthquake occurrences characterizes the probability distribution. However, a convex distribution approximated by two power-law curves is found for the cluster of all earthquakes.

Another example in Fig. 8-13 is the probability of time-intervals for an earthquake swarm of more than 2000 local earthquakes in Izu peninsula, Japan from July to December of 1989. This also follows power-law distributions rather than the exponential function of Poisson distribution. Power coefficients of d_1 and d_2 in Fig. 8-13 are about 0.5 and a little smaller than 1.0, respectively. This swarm activity has a b-value for Gutenberg and Richter's empirical relation of about 0.82, while the b-value of the earthquake activity in the vicinity of Japan shown in Fig. 8-12 is about 0.85. The convex distribution in Figs. 8-12 and 8-13 will be investigated in relation to the non-linear scaling law later.

No previous theory has established a connection between Poisson and power-law distributions for earthquake occurrence. The theory developed in Appendix A, from (A-7) to (A-14), provides the relation between the autocovariance function and the time interval distribution of the random processes. Therefore, we are able to consider jointly the long-tail behavior of the earthquake autocovariance in Fig. 8-11 and the power-law distribution of earthquake time-interval in Fig. 8-12. Note that the power coefficient of $2 - 2H$ with $H=0.6$ yields a power coefficient d of 0.8, consistent with the power coefficient of 0.5 to 1.0 for the earthquake swarm in Fig. 8-13.

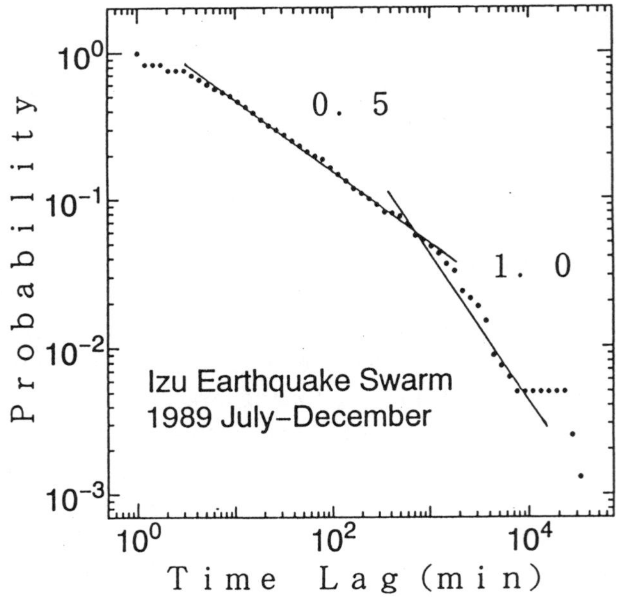

Figure 8 – 13. **Cumulative number of time intervals of successive earthquakes for the swarm activity in Izu peninsula of 1989. The value is normalized to represent the probability distribution. Two straight lines indicate the power-law dependence between cumulative number and time interval in minutes.**

The stochastic scaling in the previous section provides a clear method for reducing the family of exponential functions to a power-law function. This encourages us to develop further stochastic scaling for understanding the complexity of earthquake activity. Suppose that the probability density of successive events within a time interval from τ to $\tau + d\tau$ is $P_0(\tau)d\tau$, then the Poisson process provides

$$P_0(\tau) = \lambda_0 \exp(-\lambda_0\ \tau) \quad (\tau \geq 0), \tag{8 – 19}$$

where λ_0 is the average number of events in unit time. Let us consider that equation (8-19) represents the probability density of one series of earthquake occurrences classified by the minimum magnitude range. Here we introduce a characteristic time for this activity as

$$\tau_0 = \lambda_0^{-1}. \tag{8 – 20}$$

Another series of earthquakes classified by a slightly larger magnitude-range than that in (8-19) can be considered. The relative probability of this series of earthquake activity can be expressed similarly to the stochastic scaling of (8-13) as

$$f_s : \frac{a_s}{b_s} P_i(b_s \tau) = P_{i+1}(\tau), \qquad (8-21)$$

where a_s and b_s are positive scaling parameters. This scaling is also stochastic, because the functions which scale are the probability functions. Within the time-lag interval from τ to $\tau + d\tau$, the i-th series of earthquakes has a probability density of P_i defined relatively to P_0 of the 0-th series in (8-19). Since the scaling parameters are smaller than unity in (8-21), this scaling indicates coarse graining for earthquake activities characterized by larger magnitudes and smaller number of events. For a sufficient number of events, there is no preference for any series of events. Sometime, an event of the $(i+1)$-st series will occur, and at other times one of the $(i+2)$-nd series will occur. Since each series of earthquake activity is independent, the chance of an event occurring does not depend on the probability for the other series of earthquakes. Therefore, the chance of an event from the cluster of these series is

$$P_s(\tau) = \sum_{i=0}^{M} P_i(\tau), \qquad (8-22)$$

where $P_i(\tau)$ can be obtained formally from (8-19) and (8-21) as

$$P_i(\tau) = \lambda_0 \left(\frac{a_s}{b_s}\right)^i \exp(-b_s{}^i \lambda_0 \tau) \quad (0 \leq \tau). \qquad (8-23)$$

A characteristic time can be also defined for each series of the activity as

$$\tau_i = (\lambda_0 b_s{}^i)^{-1}. \qquad (8-24)$$

Since the total number of events classified into each series decreases as the magnitude increases, we have a constraint after integrating $P_i(\tau)$ in (8-23) with respect to τ from 0 to ∞ as

$$\frac{a_s}{b_s{}^2} \leq 1. \qquad (8-25)$$

Clearly, the above ratio relates to the scaling parameters a_d and b_d in (8-13) and the empirical b-value in (8-2) as

$$\frac{a_s}{b_s{}^2} = \frac{a_d}{b_d} = 10^{-b\Delta M}, \qquad (8-26)$$

where ΔM is the magnitude interval. This relation provides an understanding of the physical meaning of the empirical b-value in terms of the stochastic scaling in (8-21).

An asymptotic form of (8-22) can be derived similarly by the steepest descent method

$$P_s(\tau) \simeq A_s|\lambda_0 \; \tau|^{-\xi_s+1}, \tag{8 - 27}$$

$$A_s = \lambda_0 \sqrt{\frac{\pi}{2\ln(\frac{a_s}{b_s})\ln(b_s)}} \exp([\xi_s - 1][\ln(\xi_s - 1) - 1]), \tag{8 - 28}$$

where ξ_s, a scaling dimension of the stochastic scaling in (8-21), is defined as

$$\xi_s = \frac{\ln(a_s)}{\ln(b_s)}. \tag{8 - 29}$$

Inequality of (8-25) restricts this scaling dimension $\xi_s < 2$. Although the asymptotic form of (8-27) is valid for the case of $\xi_s > 1$, the power-law can be also extended for the case of $1 > \xi_s > 0$ similarly to the fractional power spectrum discussed in §7. Consequently, although each series of earthquake activity, classified by a particular magnitude-range, is characterized by a Poisson process, the cluster of many of these series exhibits a power law distribution of the autocovariance and time-interval. Therefore, the power-coefficient of d_1 and d_2 empirically obtained from cumulative number is formally understood in relation to the scaling dimension ξ_s of the complex earthquake activity as

$$d = 2 - \xi_s. \tag{8 - 30}$$

Once we determine the power coefficient of the time-interval distribution of an earthquake activity, the coefficient indicates the scaling property through (8-30). We conclude here that the fundamental property of complex earthquake activity can be simply described by the stochastic scaling. This result has been derived in a general manner, so we expect that this theory describes not only the long-tail behavior of the complicated earthquake activity but also the complex system of natural phenomena.

8.6. MAXIMUM ENTROPY OF EARTHQUAKE ACTIVITY

Empirical values of p, b, power-coefficient d's and H have been evaluated for natural earthquakes, and equation (8-26) relates the scaling parameters and b-value. Therefore, all the scaling parameters a_d, b_d, a_s, and b_s can be determined for each earthquake activity. The b-value is usually about 0.8 to 1.0 and the coefficient p is about 1 to 2 for most of these irrespective

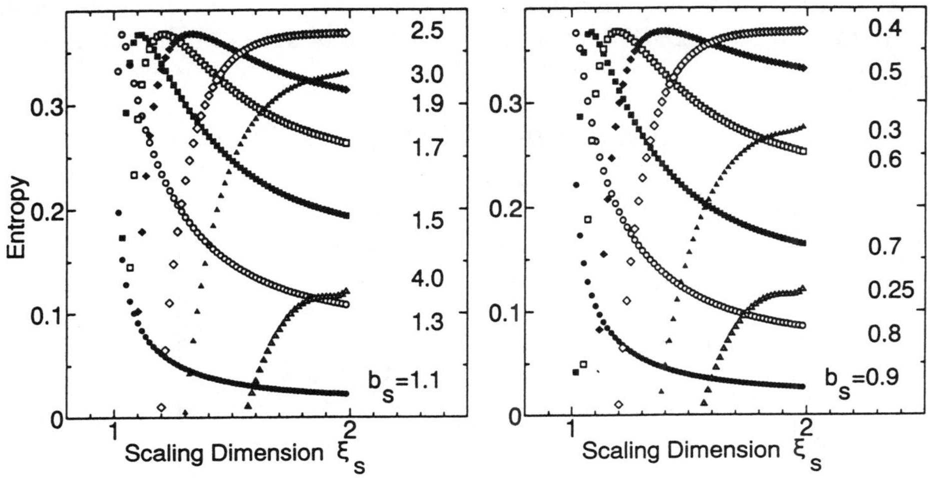

Figure 8 – 14. **Entropy of complex earthquake activity in terms of the scaling dimension ξ_s. The scaling parameter b_s is also shown.**

of the size and energy of the activity. Here we investigate how these values are preferentially determined in nature.

Since the asymptotic form in (8-27) is for $\lambda_0\tau \gg 1$, the time interval of $0 \leq \tau < \tau_0 (= \lambda_0{}^{-1})$ is not important for describing the complexity of the earthquake activity. This time interval is dependent mostly on the initial condition, when $a_s < b_s < 1$. In the time interval of $(\lambda_0 b_s{}^M)^{-1} < \tau$, there is low probability for events, because of the small number of earthquake nuclei remaining. This introduces another time constant of $\tau_M = (\lambda_0 b_s{}^M)^{-1}$. These two characteristic times, τ_0 and τ_M specify the scaling region for the complex earthquake activity. When $a_s > b_s > 1$, the time interval $\tau_M < \tau < \tau_0$ is the scaling region.

All the chances (probability) within this scaling region of time are calculated as

$$P_T = \int_{\tau_0}^{\tau_M} P_s(\tau) d\tau \bigg/ \int_0^{\infty} P_s(\tau) d\tau. \qquad (8-31)$$

The entropy of this cluster of the earthquake activity is then defined as

$$E_T = -P_T \ln P_T. \qquad (8-32)$$

Under the condition in (8-25), we can calculate E_T for each pair of a_s and b_s. Figure 8-14(a) shows the entropy (8-32) in terms of the scaling dimension

ξ_s where the scaling parameters are $1 > a_s > b_s$ and (b) for $1 < b_s < a_s$. There are three cases in Fig. 8-14 for which the entropy becomes maximum with respect to the scaling dimension: The first case is for $\xi_s \to 1$ and b_s tends to 1.0. The second case is for ξ_s about 1 to 1.4, and the third is for $\xi_s \to 2$.

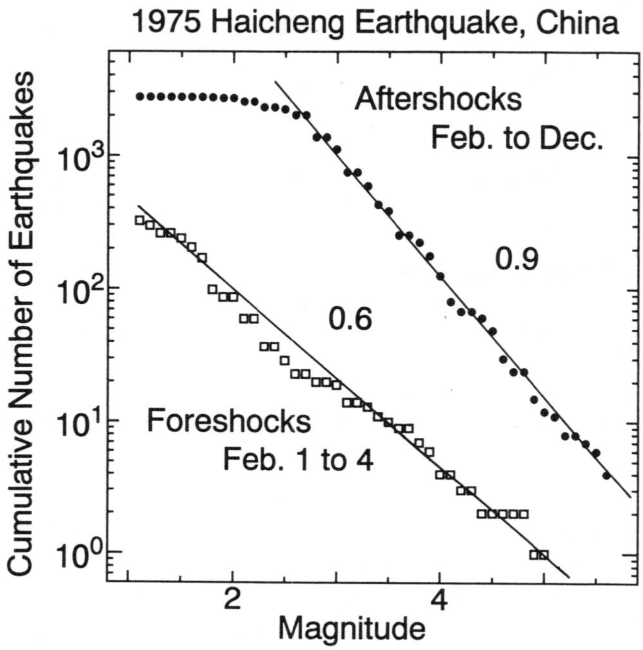

Figure 8 – 15. Cumulative number of foreshocks and aftershocks of the 1975 Haicheng earthquake in China as a function of magnitude. Empirical estimate of the b value is about 0.6 for foreshocks and about 0.9 for aftershocks (Courtesy of State Seismological Bureau of China).

The first case of the maximum entropy can be readily understood. Since the scaling parameter a_s also tends to 1.0 in this case, this is just the case without the scaling. All the series comprising the cluster of activity are the same. Therefore, each series in this case occurs at random with an equal probability, and obviously the entropy of such activity is maximum. Meanwhile, the third case indicates equal probability of series of activity, even though the component series are scaled by the scaling parameters a_s and b_s. Since the latter case manifests $\frac{a_s}{{b_s}^2} \to 1$ instead of the inequality in (8-25), the probability and/or the total number of events of each cluster is the same. It is also reasonable that this case corresponds to the maximum entropy. In the second case, the complexity of the earthquake activity gives rise to the maximum entropy.

Since an earthquake swarm occurs in a limited source area, the magnitude range is usually small. Therefore, the summation of the series of earthquake activities covers a narrow range of magnitudes. This corresponds to an extreme case without the scaling, which yields a power-coefficient d of about 1.0 and $\xi_s \rightarrow 1$. This is the first case for the maximum entropy of complex earthquake activity.

For a typical value of $b = 0.8$ with $\Delta M = 0.2$ and power-coefficient $d = 0.8$ for world-wide seismicity, the scaling dimension $\xi_s (= 2 - d)$ is calculated to be 1.2. In this case we observe a_s of about 1.74 and b_s of about 1.58 from (8-26) and (8-30). These values are almost the same as those for the earthquake swarm in Fig. 8-13, where d of $2 - \xi_s$ is about 0.8. Suppose that b_s is 1.6, the theoretical scaling dimension for the second case of the maximum entropy is about 1.2 in Fig. 8-14. This indicates that the world-wide seismicity in Fig. 8-11 and the earthquake swarm in Fig. 8-13 are well described by the maximum entropy complex activity.

Although the earthquake activity is complex, it is typically represented by the above b- and d-values which are expressed by the physical scaling parameters satisfying the condition of the maximum entropy. Note that the characteristic time of larger earthquakes is shorter than that of smaller earthquakes, since $b_s > 1$ in this analysis. This may seem contradictory to our intuition, but this suggests that large aftershocks seldom occur and most of the aftershocks are very small long after the main shock.

For p of 1.1 and b of 0.8, for example, the fractal dimension of ξ_d is 1.1. In this case a_d and b_d are about 0.017 and 0.025, respectively. For p of 1.05 which has been calculated from the data in Fig. 8-1, we observe a_d of 0.0004 and b_d of about 0.0006. The scaling parameter b_d becomes smaller as $p \rightarrow 1$. Since b_d measures the time unit, this gives an extremely long time-constant for decaying aftershocks. This is the reason why the aftershocks of the 1891 Nobi earthquake last a very long period of time, since p of this earthquake is very close 1. From the relationship of $p = \xi_d$ with (8-26), we can show that scaling dimension ξ_s tends to 2 as $p \rightarrow 1$. Therefore, the long-tail behavior of aftershock activity corresponds to the third case of the maximum entropy.

Foreshock activities with b-values as small as 0.5 have been found. In case of the Haicheng earthquake of 1975 in China, for example, b value of the foreshocks has been measured at about 0.56 (Fig. 8-15). Figure 8-16 shows the time interval distribution of foreshocks and aftershocks of the 1975 Haicheng earthquake. Although the power-coefficient d of aftershocks is not well determined, d of about 0.8 is obtained for the time interval distribution of foreshocks. The second case of the maximum entropy in Fig. 8-14 with $b = 0.6$ and $d = 0.9$ predicts ξ_s of about 1.1 and b_s of about 1.4. When we compare these theoretical values with the above observation,

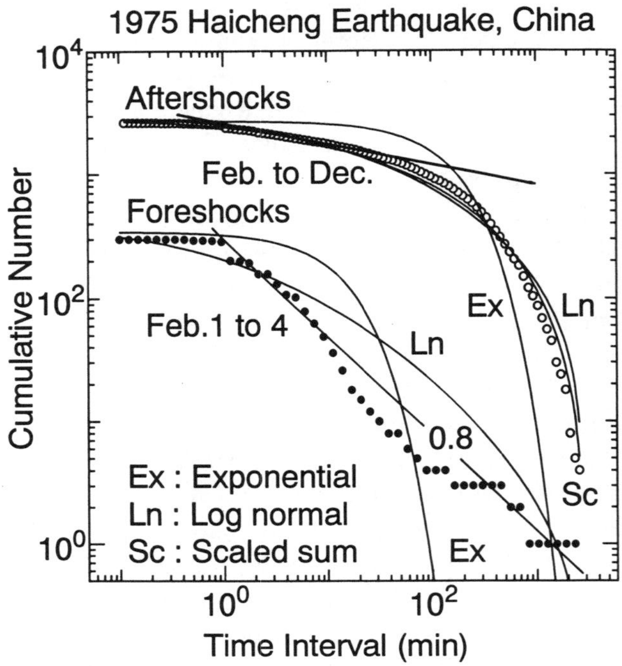

Figure 8 – 16. Cumulative number of foreshocks and aftershocks of the 1975 Haicheng earthquake in China as a function of successive time interval. Solid lines indicate a power law with coefficient of about 0.8 for foreshocks and about 0.16 for aftershocks. Solid curves in the figure indicate Poisson and log-normal distributions for foreshocks and aftershocks. Scaled-sum distribution for aftershocks is also shown.

the empirical values of b and d in Figs. 8-15 and 8-16 is consistent with the theoretical prediction. This is very important for understanding the essential property of foreshocks to predict following major earthquakes.

8.7. FURTHER APPLICATION OF STOCHASTIC SCALING

The power-law with a coefficient of about 0.8 fits the time-interval distribution of successive foreshocks of the 1975 Haicheng earthquake (Fig. 8-16) but the exponential distribution would not. The distribution function of aftershocks is not well determined, since the decay rates of medium and large time-intervals are very small compared with the exponential but not as small as the decay of a power-law. A common analysis for such asymmetric data is to apply a log-normal distribution, since the distribution has both a long-tail and statistical moments. Log-normal distributions have been observed in many diverse fields, and the time-interval distribution of successive aftershocks with variety of sizes seems to be another example as shown in Fig.8-16. We reconsider this asymmetric distribution from the viewpoint of our stochastic scaling.

If $\ln x$ has a normal distribution then the variable x has the distribution

$$g\left(\frac{x}{\bar{x}}\right)\frac{dx}{\bar{x}} = \frac{1}{\sqrt{2\pi\sigma^2}}\exp\left(-\frac{[\ln(x/\bar{x})]^2}{2\sigma^2}\right)\frac{dx/\bar{x}}{x/\bar{x}}. \qquad (8-33)$$

When the variance σ^2 is large and/or $x \simeq \bar{x}$, $g(x)$ mimics a $1/x$ distribution (Montroll and Shlesinger, 1983). The larger σ, the more orders of magnitude the mimicking persists. In the meantime, the $1/x$ distribution can be derived as an asymptote (8-27) when $\alpha_s \to {\beta_s}^2$. Therefore, the log-normal distribution can be approximated by our stochastic scaling method of (8-22). Provided that $\alpha_s = {\beta_s}^2$, β_s (< 1.0) is the only parameter. Scaled-sum distribution has been calculated taking different values of variance σ^2 in (8-33), and two probability distributions were compared each other. The numerical result confirms the consistency between these two distributions for the probability range of 99%. The rest of 1% is quit different because of the difference in functional forms.

The concaved distribution of regional earthquakes in the vicinity of Japan (Fig.8-12) and that of the Haicheng aftershocks (Fig.8-16) is so understood by the stochastic scaling of the cluster of earthquake activities. The result is shown in Fig.8-16, resolving the consistency of log-normal and stochastic-scaling distributions fitted to the data. This is also characterized by the third case of the maximum entropy, since the scaling dimension ξ_s tends to 2. This conclusion has been derived in a general manner, so that the present approach can be applied to not only the long-tail behavior of the complex earthquake activity but also to the complex system of natural phenomena described by the log-normal distribution.

8.8. DISCUSSIONS ON COMPLEX EARTHQUAKE ACTIVITY

Various time sequences of the complex earthquake activity have been studied by introducing a simple mathematical model. The basic principle is to apply the stochastic scaling to a cluster of many series of earthquakes classified by specific magnitude (energy) ranges. Each series is modeled by a stochastic process of random earthquake occurrences. The stochastic scaling describes the mutual relation among statistical properties of the series. Therefore, the complexity of the earthquake activity is represented by scaling parameters and the total number (or total energy) of the activity. Thus the physical basis of this model is quite simple and far-reaching. Since the stochastic scaling is considered in the fundamental equation without any externally-imposed stochasticity or heterogeneity, we are able to evaluate the total energy and the entropy of the complex system of earthquake occurrences. Without this basic approach, we cannot determine how the scaling parameters are preferentially determined in nature.

Since the earthquake swarm occurs in a limited source area, the magnitude range of earthquakes is usually very narrow. Therefore, the stochastic size-effect is not essential to explain the temporal variation of the earthquake swarm activity. This reduces the temporal variation to the production and reduction of latent sources of earthquakes in the crust and to the exponential decay of the activity. This is important in understanding the complex earthquake activity, because the entropy of this type of swarm activity is maximum.

Magnitudes of aftershocks of large earthquakes cover a wide range. It is reasonable to apply the asymptotic form of (8-15) in such size-dependent random phenomena. Consequently, the power-law decay of aftershock numbers empirically derived is understood in terms of the complexity of the earthquake activity. The stochastic scaling here predicts the Omori-Utsu empirical constant to be in the range of $1 < p < 2$. For $p < 1$ the stochastic scaling does not apply. The long-tail of aftershock activity is observed when the empirical power coefficient p is about 1.0 and the activity is also characterized by the maximum entropy. The distribution described by the log-normal is also in this category.

The present approach differs from previous analyses dealing with non-linear dynamical systems to simulate the earthquake occurrence in the manner above. Even for a few degrees of freedom, non-linear dynamical systems exhibit chaotic behavior showing complicated patterns of earthquake-like occurrences. Previous models predict a size-number distribution of random phenomena similar to what is seen in the earthquake activity. However, the random phenomena from these model simulations are not always applicable to earthquake phenomenon. For example, the slip velocity on a fault plane can be evaluated as a random variable by the numerical simulations of the previous models of the self-organized criticality (e.g., Bak and Tang, 1989; Carlson, 1991). Slip velocity varies over many orders of magnitude, because it is a random variable. Slip velocity on the natural faults on the other hand is almost constant and does not change by as much as a factor of 2. This is an important constraint for understanding the kinematical similarity in §2.

Obviously, the basic features which are observed in the previous models should be representative of a wide class of physical models. These provide clues for understanding the generation mechanism of the earthquake phenomenon, one of which has been studied in §8.2 and §8.3. Nevertheless, further insight is needed before comparing directly the results obtained from the models with those of natural earthquakes.

The strength and the number of earthquakes are statistically scaled in the work here. We have not discussed the physical origin of the scaling, however the scaling property can be found in the diversity of the complex systems in nature. The stochastic scaling here is one such scaling representing a fundamental aspect of the complex system. As a consequence of this scaling, we understand earthquake swarms as rupture processes of fault patches without the scaling or with a characteristic short-wavelength. This is consistent with the simulation showing that a highly inhomogeneous stress field with short-wavelengths increases the frequency of earthquake occurrence. Since the stochastic scaling performs a coarse-graining of statistical properties of earthquake occurrences, an earthquake sequence of foreshocks, mainshocks and aftershocks is the result of random occurrences of events, which constitutes the local and global seismicities.

CHAPTER 9

NON-LINEAR SCALING LAW OF EARTHQUAKE ACTIVITY

In addition to the stochastic scaling in the previous section, we will further investigate the hierarchy of the complex system. In order to unify the local, regional and global seismicities, a non-linear scaling law is derived. The non-linear scaling law characterizes the hierarchy of the complex system as well as earthquake activity in a general manner.

The autocovariance of earthquakes in and near Japan is represented by a power-law function which is identical to that of world-wide seismicity (Ogata and Abe, 1991). The time interval distribution of an earthquake swarm in Izu and of major earthquakes in and near Japan is also represented by the power law. Power-law distributions for local and regional earthquake activity do not necessarily require a power-law distribution for the global earthquake activity. Figure 9-1 shows the Gutenberg and Richter's relation of shallow moonquakes (Nakamura, 1980). The decay rate (b value) in the large magnitude range is larger than that in the small magnitude range. The faster decay rate in large magnitude range has been also found in b values of natural earthquakes (Pacheco et al., 1992). Gutenberg and Richter's empirical relations for local earthquakes, regional earthquakes and global earthquakes have been obtained and they suggest the power law in variety of earthquake source sizes. This is another example indicating the hierarchy represented by the power law, where the component subsets are also characterized by power laws.

Figure 9-2 (a) shows the vascular system on a cat brain surface. Also shown is the number of boxes $N(s)$ which includes tracts in terms of grid size s for the box counting method[§]. Power laws of $N(s) \propto s^{-1.30}$ and $N(s) \propto s^{-1.77}$ represent the distributions in short and long grid-ranges. Longer grid-range than 0.5 mm, the power coefficient is about -2. Figure 9-2 (b) shows those of one particular portion of the whole. The above power-law also explains this subset distribution, suggesting the self-similarity of the vascular system.

These observations indicate that there is some universal nature which connects power-law distributions of local (subset) activities with that of the global (full set) activity. Hara and Koyama (1992) investigated the scaling relation between local and global complex systems in a general manner. We slightly modify the original theory and apply it to the complex earthquake activity in this chapter.

[§]Consider square boxes of the grid $s \times s$, and count the number $N(s)$ of such square boxes required to cover the trace of a curve on a plane. $N(s) \times s$ is expected to approach a constant value of the total length of the curve. However, this is not the case.

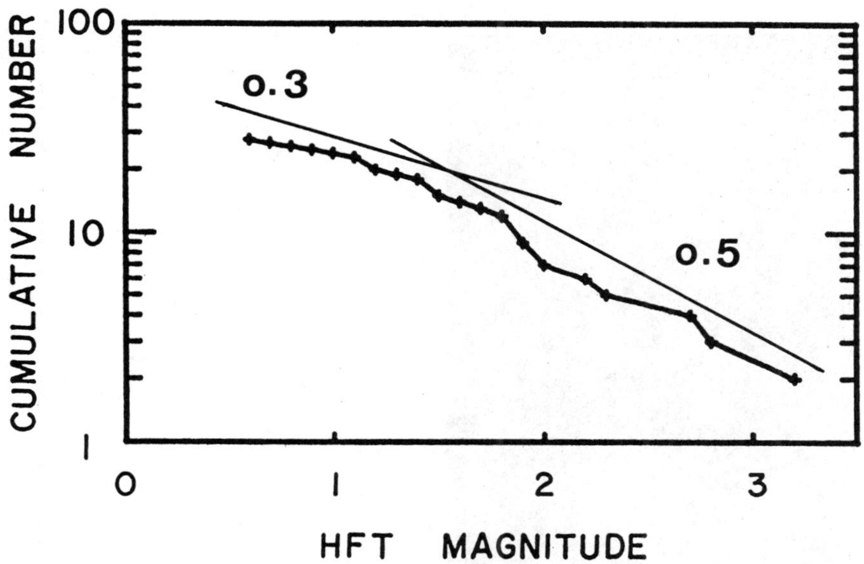

Figure 9 – 1. Magnitude-frequency distribution of observed shallow moonquakes. HFT magnitude is defined by logarithm of the maximum envelope-amplitude recorded by short-period lunar seismometer. The value is estimated to be about 1.5 less than the Richter magnitude (Nakamura, 1980).

9.1. HIERARCHY OF EARTHQUAKE ACTIVITY

Suppose that there is a cluster of earthquake activity represented by a power-law autocovariance as in Fig. 8-11,

$$C_j(\tau) \propto \tau^{-\eta_j}. \qquad (9-1)$$

For a global standpoint, there exist w_j of such clusters characterized by the same value of η_j. Because the global activity is a summation of many different clusters and because clusters are considered to be independent, the global nature is expressed by the sum of (9-1) as

$$C(\tau) = \sum_{j=0}^{M} w_j \; C_j(\tau). \qquad (9-2)$$

The above summation means in other words that there is no interaction among different clusters occurring in different places. We contrast this with the independency of the series of activities within one cluster. The functional form of $C(\tau)$ must be

$$C(\tau) \propto \tau^{-q}, \qquad (9-3)$$

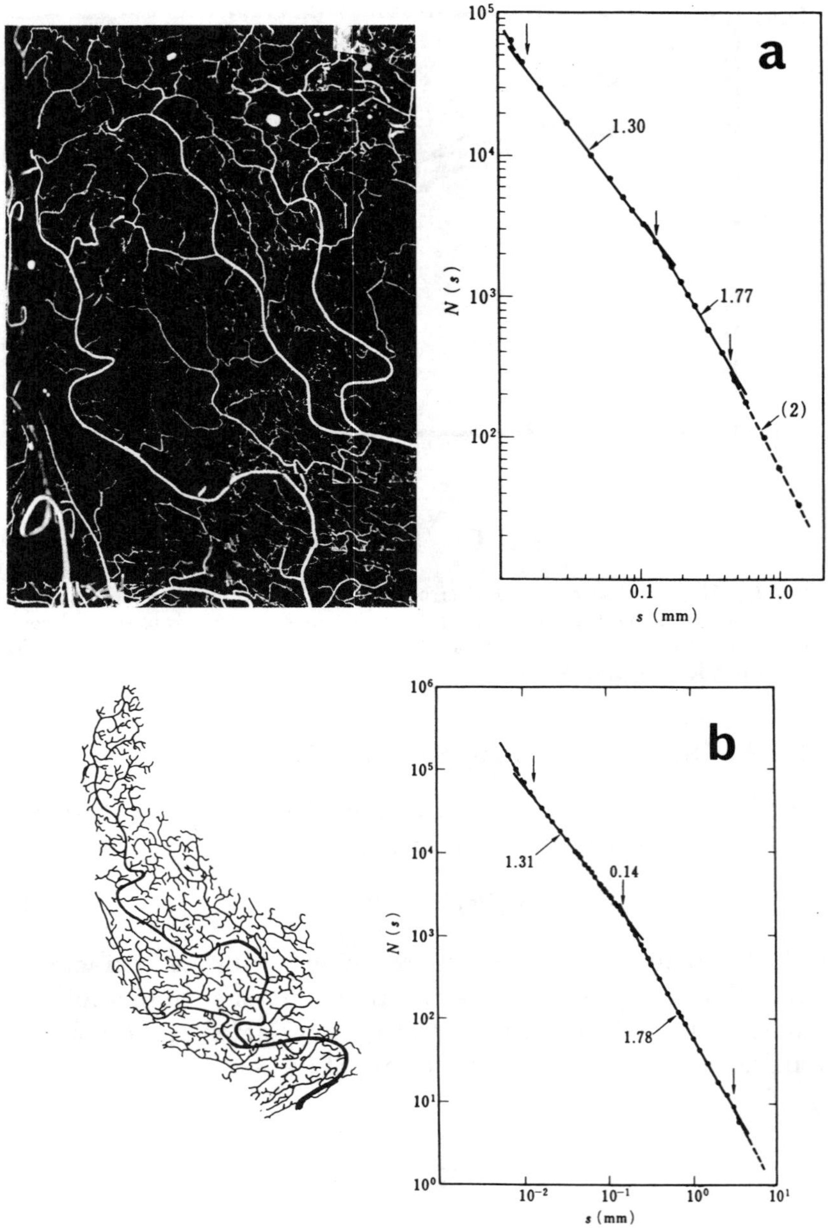

Figure 9 – 2. Fractal analysis of the vascular system. (a) Trace of vascular tracts on a cat brain surface (left). Box number $N(s)$ against box grid size s of the tract distribution is shown in the right. Power-laws in short, long and much longer than 0.5 mm ranges are indicated by the power coefficient of -1.30, -1.77 and -2. (b) The same for one artery vessel on the left. The power law of this subset vascular system is consistent with the whole (Matsuo et al., 1990).

in order to explain the power-law autocovariance of the global earthquake activity, where q is a constant or may weakly depend on τ. This is what we understand from empirical power-law distributions for local and global earthquake activities, where η_j and q can be determined empirically.

In order to evaluate (9-2) and to clarify the relation in terms of w_j, the clusters are re-labeled in increasing order of η_j as $0 \leq \eta_1 < \eta_2 < ... < \eta_M$. Taking a new variable x of $\eta = \eta_M\ x$ $(0 \leq x \leq 1)$, a formal representation of (9-2) and (9-3) is given as

$$\int_0^1 \tau^{-\eta_M\ x} w(x)dx \simeq \Psi\ \tau^{-q}, \qquad (9-4)$$

where Ψ indicates the level of the global earthquake activity. The weighting function $w(\tau)$ satisfies a relation

$$w(x) = \tau^{-q}\ \eta_M \ln\tau\ w(\eta_M\ x \ln\tau). \qquad (9-5)$$

Since this equation represents the scaling relation of $w(x)$, it leads to

$$w(x) \propto x^{\kappa(\tau)}. \qquad (9-6)$$

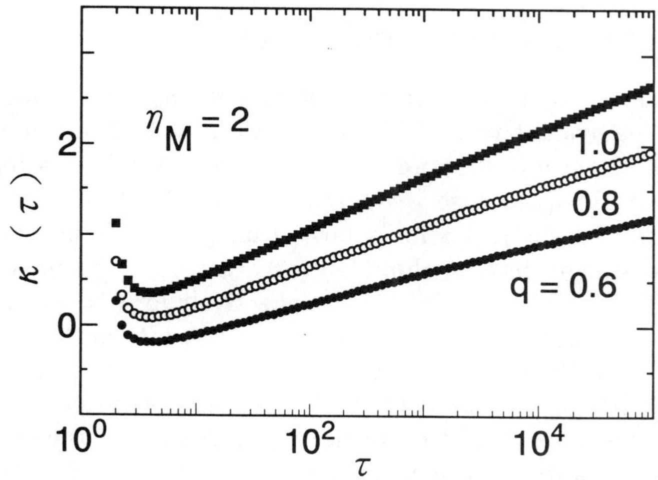

Figure 9 – 3. Power coefficient $\kappa(\tau)$ of the non-linear scaling law in terms of time-lag τ.

The integrand of the left-hand side of (9-4) has a peak value for $x = \dfrac{\kappa(\tau)}{\eta_M \ln\tau}$ in the range $0 \leq x < 1$. The integrand decreases monotonically for

$x > \dfrac{\kappa(\tau)}{\eta_M \ln \tau}$. It is negligibly small for $x > 1$, so the integral in (9-4) can be extended approximately to infinity without losing generality. Then, (9-4) is

$$\begin{aligned}
&\int_0^\infty \exp(-\eta_M\ x \ln \tau)\ x^{\kappa(\tau)} dx \\
&= \int_0^\infty \exp(-\eta_M\ x \ln \tau)\ \tau^{-q} \eta_M \ln \tau\ w(\eta_M\ x \ln \tau) dx \\
&= \tau^{-q} \int_0^\infty \exp(-X)\ w(X) dX, \qquad (9-7)
\end{aligned}$$

where

$$X = \eta_M\ x \ln \tau, \qquad (9-8)$$

and

$$\kappa(\tau) = \frac{q \ln \tau}{\ln |\eta_M \ln \tau|} - 1. \qquad (9-9)$$

The last integral of the right-hand side of (9-7) changes little with respect to τ. Therefore a non-linear weighting function of $w(x) \propto x^{\kappa(\tau)}$ provides the theoretical connection between the power law for local activity and that for global activity.

9.2. NON-LINEAR SCALING OF COMPLEX SYSTEM

Figure 9-3 shows $\kappa(\tau)$ as a function of τ. Here η_M is assumed to be 2, because η is equivalent to $2 - 2H$ with $0 < H \leq 1$. The power constant q is assumed to be 0.6, 0.8 and 1.0. $\kappa(\tau)$ is a slowly-increasing function of τ. For $q = 0.8$ (the case in Ogata and Abe (1991)), $\kappa(\tau)$ increases from about 0.7 to about 1.5 for a large change in τ from 100 to 10000. In a large lag-range, $\ln \tau$ of about 10 for example, $C(\tau)$ is composed of $0.689\tau^{-1.6}$, $0.216\tau^{-0.8}$ and $0.068\tau^{-0.4}$, respectively for $x = 0.8$, 0.4 and 0.2. The density of faster decay of $\tau^{-1.6}$ is ten times larger than that of the slower decay of $\tau^{-0.4}$. While in a small lag-range, $\ln \tau$ of about 5 for example, $C(\tau)$ of $0.848\tau^{-1.6}$, $0.509\tau^{-0.8}$ and $0.305\tau^{-0.4}$. The density difference in this case is less than a factor of 3. These calculations suggest that the larger the lag is, the faster the autocovariance decays. This is the explanation for what we have seen in Fig. 8-13 from the view point of the scaling of complex system.

The stochastic scaling in (8-13) or (8-21) describes the hierarchy of a cluster of many series of local earthquake occurrences, and is described by a linear mapping function. Therefore, such the stochastic scaling applies to the local earthquake activity. However, the scaling of many of these clusters cannot be expressed by a simple linear function. Although we could not specify the mapping function, the scaling involved with the weighting function of $w(x)$ is highly non-linear. This non-linear scaling applies to the global earthquake activity. Consequently, the stochastic scaling and the non-linear scaling describe the hierarchy of the complex earthquake activity on both local and global scales.

This non-linear scaling is different from the simple hierarchy of the complex system. The family of random phenomena characterized by exponential functions leads to the long-tail or slow dynamics represented by power-law functions. For this we have studied the stochastic scaling. The family of slow dynamics by power-laws is expected to be super-slow dynamics characterized by $(\ln t)^{-\zeta}$ (Hara et al., 1996). This can be understood as following relationships:

$$\sum \left(\frac{a}{b}\right)^j \exp(-b^j \bar{\gamma} t) \rightarrow t^{-\kappa},$$

$$\sum \left(\frac{a'}{b'}\right)^j \exp(-b'^j \kappa \ln t) \rightarrow (\ln t)^{-\zeta},$$

because of $t^{-\kappa} = \exp(-\kappa \ln t)$. Therefore, super-slow dynamics constitutes the hierarchy of complex systems, while the non-linear scaling law characterizes the composite system for the cluster of many complex systems.

The long-tail behavior of the local, regional and global seismicities is the evidence of the hierarchical structure of earthquake occurrence. The essential property of the hierarchy is described by the non-linear scaling law. The present theory has been applied to the autocovariance function, and it can be applied to any function represented by the power law. The theory is also applicable to Gutenberg and Richter's relations for small earthquakes and for great earthquakes in the world. Therefore, the b-value variation of natural earthquakes in Fig. 9-1 and the size-distribution of vascular system in Fig. 9-2 can be similarly understood. The non-linear scaling law explains the variation in the power coefficients of the complex system characterized by the power-law.

Appendix A: Autocovariance of Dislocation Velocity Function

When we consider a random dislocation velocity function in Fig. 2-2(b), one of its sample autocovariance function is

$$C(\tau) = \begin{cases} \frac{1}{T_0}\int_0^{T_0-|\tau|} \dot{D}(\xi,t)\dot{D}(\xi,t+|\tau|)dt, & 0 < |\tau| \leq T_0; \\ 0, & T_0 < |\tau|. \end{cases} \qquad (A-1)$$

A realization of a stationary stochastic process $\dot{D}(\xi,t)$ can be considered as $T_0 \to \infty$. It has an autocovariance function

$$\begin{aligned} R(\tau) &= \lim_{T_0\to\infty} \frac{1}{2T_0}\int_{-T_0}^{T_0} \dot{D}(\xi,t)\dot{D}(\xi,t+\tau)dt \\ &= \bar{a}^2 + \sigma^2 \exp(-\lambda|\tau|). \end{aligned} \qquad (A-2)$$

This will be given at the end of this appendix. The estimator of autocovariance function (A-1) is

$$\begin{aligned} E[C(\tau)] &= E\Big[\frac{1}{T_0}\int_0^{T_0-|\tau|} \dot{D}(\xi,t)\dot{D}(\xi,t+\tau)dt\Big] \\ &= \frac{1}{T_0}\int_0^{T_0-|\tau|} R(\tau)dt \\ &= (1-\frac{|\tau|}{T_0})\{\bar{a}^2 + \sigma^2 \exp(-\lambda|\tau|)\}, \end{aligned} \qquad (A-3)$$

where E stands for the expectation of many samples of $C(\tau)$, because of the ergodicity of the stationary stochastic process $\dot{D}(\xi,t)$.

Similarly, an estimator of autocovariance function along the fault length is

$$E\Big[\frac{1}{L}\int_0^{L-|\eta|} d\xi \ \frac{1}{T_0}\int_0^{T_0-|\tau|} dt\dot{D}(\xi,t)\dot{D}(\xi+|\eta|,t+|\tau|)\Big]. \qquad (A-4)$$

Since all the statistical properties of the stationary stochastic process of $\dot{D}(\xi,t)$ are the same for any sample of realizations due to the ergodicity of $\dot{D}(\xi,t)$, the expectation (A-4) satisfies

$$E\Big[\int \dot{D}(\xi,t)\dot{D}(\xi+\eta,t)d\xi\Big] = E\Big[\int \dot{D}(\xi,t)\dot{D}(\xi,t-\frac{\eta}{\bar{v}})d\xi\Big]. \qquad (A-5)$$

This is just the equivalency between the time average and space average of the stationary stochastic process. Therefore, the estimator of (A-4) is

$$(1-\frac{|\eta|}{L})(1-\frac{|\tau-\eta/\bar{v}|}{T_0})\{\bar{a}^2+\sigma^2\exp(-\lambda|\tau-\frac{\eta}{\bar{v}}|)\}. \qquad (A-6)$$

In the text, we recover the physical dimensions of L and T_0.

In order to derive the autocovariance function (A-2), we consider a random variable $x(t)$ along the time axis. Random amplitude in Fig. 2-2 (b) could be represented in the n-th time interval

$$x(t) = a_n, \qquad t_n < t < t_{n+1} \tag{A-7}$$

where $..., a_{n-1}, a_n, a_{n+1}, ...$ form a set of mutually independent random variables. Its mean and mean square value are

$$E[x(t)] = E[a_n] = \bar{a}, \tag{A-8}$$

$$E[x(t)^2] = E[a_n^2] = \overline{a^2}. \tag{A-9}$$

The autocovariance function $R_{xx}(t_1, t_2) = E[x(t_1)x(t_2)]$ can be calculated as follows: For a Poisson distribution with average λ, the probability of an event that contains $x(t_1)$ and $x(t_2)$ in the same time interval is

$$P = \exp(-\lambda|t_2 - t_1|). \tag{A-10}$$

For $x(t_1)$ and $x(t_2)$ not in the same time interval, the probability is

$$P^* = 1 - \exp(-\lambda|t_2 - t_1|). \tag{A-11}$$

Therefore,

$$R_{xx}(\tau) = P \times E[x(t_1)x(t_2)|P] + P^* \times E[x(t_1)x(t_2)|P^*], \tag{A-12}$$

where τ is the interval of $t_2 - t_1$, and the conditional expected values of $x(t_1)x(t_2)$ are

$$\begin{aligned} E[x(t_1)x(t_2)|P] &= E[x(t)^2], \\ E[x(t_1)x(t_2)|P^*] &= E[x(t_1)]E[x(t_2)]. \end{aligned} \tag{A-13}$$

Consequently,

$$R_{xx}(\tau) = \bar{a}^2 + \sigma^2 \exp(-\lambda|\tau|), \tag{A-14}$$

where

$$\sigma^2 = \overline{a^2} - \bar{a}^2. \tag{A-15}$$

Appendix B: Approximate Spectrum for Short-period Body-waves

Suppose that a time series is generated by a convolution of random impulses $z(t)$ with an impulse response function $h(t)$. The autocorrelation function of

$z(t)$ which is composed of equally-likely positive and negative unit impulses is

$$\phi_{zz}(s) = k\delta(s), \tag{B-1}$$

where k is the average number of impulses per unit time and $\delta(s)$ is Dirac delta function. The impulse response represents the response of the crustal structure and observation system, and is assumed to be

$$h(t) = h_m \exp(-a\ t) \sin b\ t \quad (t \geq 0). \tag{B-2}$$

The character of short-period P-waves requires $b = \frac{2\pi}{t_0}$, where t_0 is the apparent period of component waves, about 1 sec. Here $a = t_0^{-1}$ is also assumed. A normalization of the function is $h_m = \exp(\frac{a\ t_0}{4})$. The impulse response has Fourier transform

$$H(\omega) = \frac{h_m b}{(a + i\omega)^2 + b^2}, \tag{B-3}$$

where ω is an angular frequency. The power spectrum of the time series of $z(t) * h(t)$ is

$$Z(\omega) = k|H(\omega)|^2. \tag{B-4}$$

Consider the case in which random impulses occur densely within a duration of $h(t)$, so that many impulse responses overlap. The average power spectrum is approximated as

$$\bar{Z}(\omega) = \frac{k\ p^2\ b^2}{|(a + i\omega)^2 + b^2|^2}, \tag{B-5}$$

where p^2 corresponds to the effective value of alternating current in unit time as

$$\begin{aligned} p^2 &= \int_0^\infty {h_m}^2 \exp(-2a\ t) \sin^2 b\ t dt \\ &= \frac{h_m^2 b^2}{4a(a^2 + b^2)}. \end{aligned} \tag{B-6}$$

The energy spectrum is obtained by applying a time window to the power spectrum. Considering a Gauss time window

$$w(t) = A \exp(-\frac{t^2}{2\tau^2}), \tag{B-7}$$

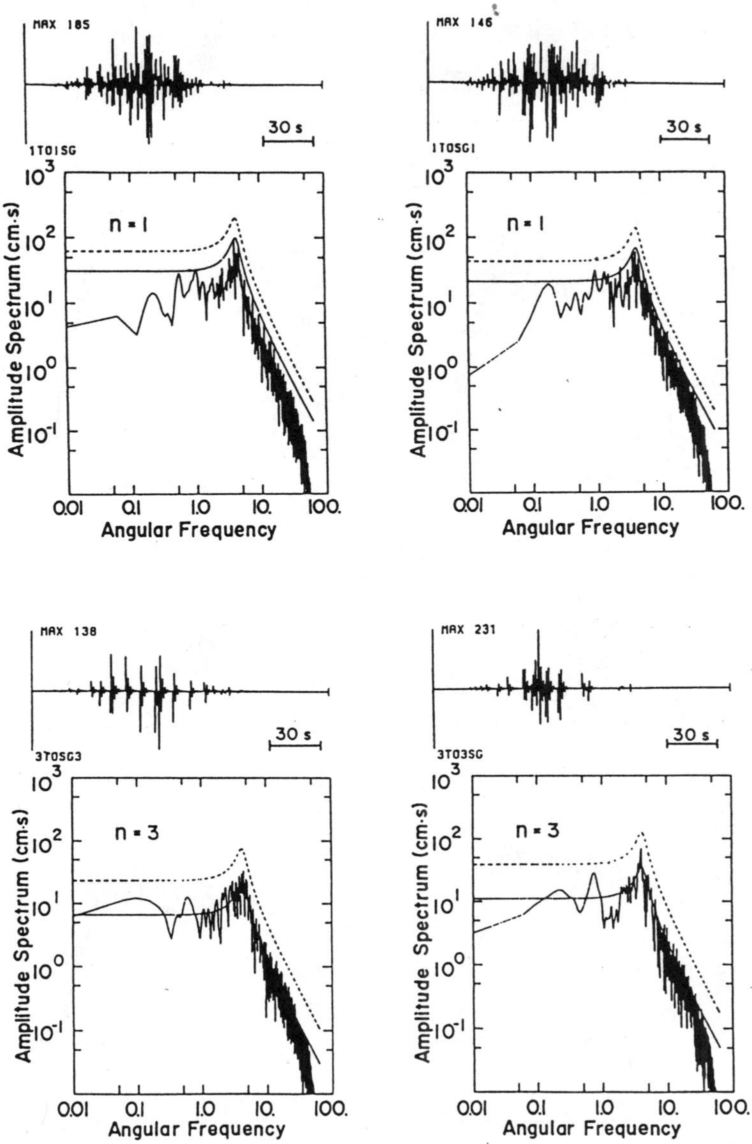

Figure B– 1. Simulation of short-period seismograms with random phase and their spectra. Synthetic seismograms are generated by a convolution of random impulses and an impulse response function. Solid curves represent spectra evaluated from (B-10) (dense case) and dashed curves from (B-11) (sparse case). Apparent period t_0 of short-period waves is assumed to be 1.5sec and effective duration of window function is 20sec. The average density of random impulses k is assumed to be $1/t_0$ and $1/3t_0$, corresponding to $n = 1$ and $n = 3$ in the figure. Spectrum of each synthetic seismogram is evaluated by FFT algorithm.

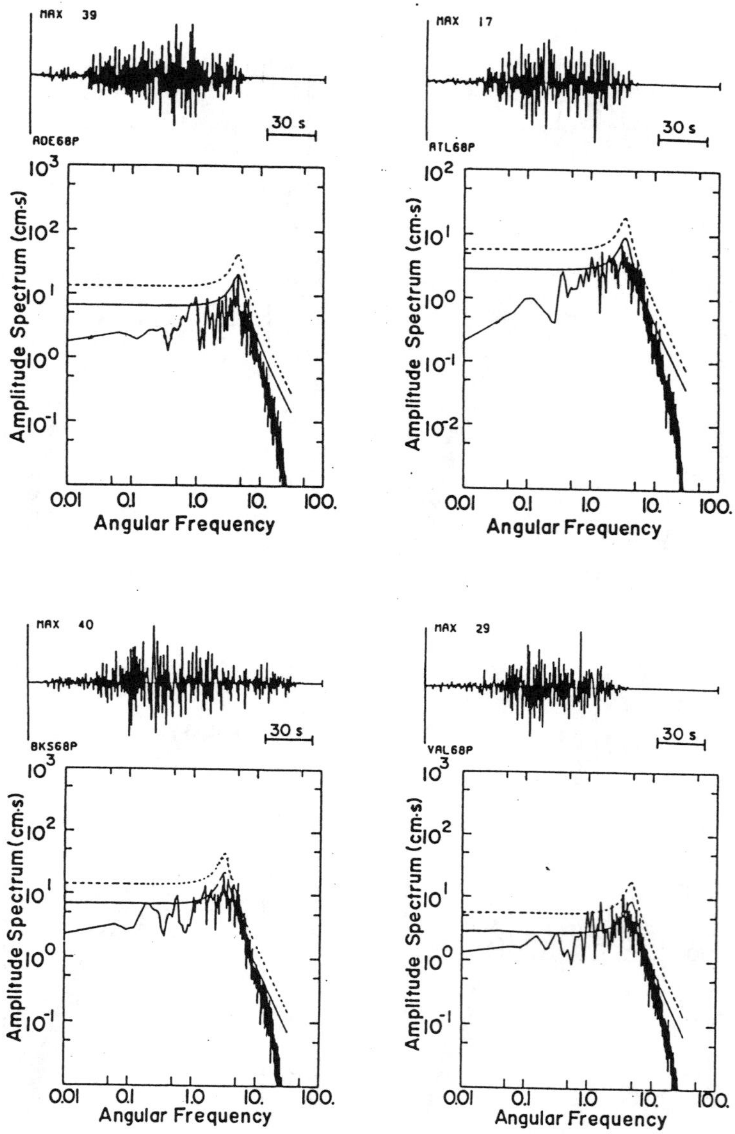

Figure B- 2. Spectra of short-period body-waves at ADE, ATL, BKS and VAL for the Tokachi-oki earthquake in Japan of May 16, 1968. The seismograms are digitized at all turning and reflection points, and interpolated at an equal time interval of 0.1 sec by 4-point Lagrange interpolation. Spectrum of each seismogram is calculated by FFT algorithm. Solid and dashed curved indicate the approximate spectra in the text.

its Fourier transform is

$$W(\omega) = \sqrt{2\pi\tau^2}A\exp(-\frac{\tau^2\omega^2}{2}). \qquad (B\text{-}\ 8)$$

Then, the energy spectrum is obtained by the convolution in the frequency domain as

$$\bar{E}(\omega) = \int_{-\infty}^{\infty} \{W(g)\}^2 \bar{Z}(\omega - g) dg. \qquad (B-9)$$

The steepest descent method is applied to evaluate the convolution in (B-9):

$$\bar{E}(\omega) = 2\pi\sqrt{\pi}\tau A^2 \frac{k\ p^2\ b^2}{|(a+i\omega)^2 + b^2|^2}, \qquad (B-10)$$

or in the case of sparse impulses

$$E(\omega) = 2\pi\sqrt{\pi}\tau A^2 \frac{k\ h_m^2\ b^2}{|(a+i\omega)^2 + b^2|^2}. \qquad (B-11)$$

The amplitude spectrum is obtained by taking the square root of (B-10) and (B-11). For a fixed number k of impulses, (B-10) and (B-11) indicate that the maximum amplitude A, build-up time τ, apparent period t_0 are the parameters needed to estimate these approximate spectra.

Figure B-1 shows synthetic random waves generated by $z(t)$ and $h(t)$. The case of $n = 1$ indicates a mean probability of random impulses, $k = \frac{1}{t_0}$, and $n = 3$ indicates $k = \frac{1}{3t_0}$. The spectrum of each synthetic random wave is plotted in the figure. Spectra for $n = 1$ are well explained by solid curves calculated from (B-10). Sparse occurrences of impulses for $n = 3$ case spectra which are similar to those from (B-11). Observed short-period P-waves are quite similar to those for $n = 1$, the highly-overlapped impulse responses. Figure B-2 shows short-period P-waves at ADE, ATL, BKS, and VAL for the Tokachi-oki earthquake in Japan of May 16th, 1968. Spectra are calculated from digitized seismograms by the FFT algorithm. Approximate spectra from (B-10) and from (B-11) with $n = 1$ are also illustrated in Fig. B-2. The data are more consistent with the approximation of (B-10) by solid curve rather than that of (B-11) by dashed curve.

Appendix C: Maximum value and *rms* value

Suppose that the probability density function of a random variable x is expressed by $P(x)$. The probability that x is smaller than a particular value of r is

$$\begin{aligned} \int_0^r P(x)dx &= 1 - \int_r^{\infty} P(x)dx \\ &= 1 - \phi(r). \end{aligned} \qquad (C-1)$$

The probability that N samples are smaller than r is $(1-\phi(r))^N$. If at least one realization exceeds the value r, its probability is

$$1-(1-\phi(r))^N. \qquad (C\text{-}2)$$

Therefore, the probability that the random variable x is from r to $r+dr$ is calculated as the probability that at least one sample of x exceeds the value of r minus the probability that at least one sample of x exceeds the value of $r+dr$:

$$\begin{aligned} &\{1-(1-\phi(r))^N\}-\{1-(1-\phi(r+dr))^N\} \\ \simeq\ & -{}_NC_1(1-\phi(r))^{N-1}\frac{d\phi}{dr}dr \\ =\ & N(1-\phi(r))^{n-1}P(r)dr. \end{aligned} \qquad (C\text{-}3)$$

The most probable value of the maximum x can be evaluated by differentiating (C-3) by r.

For a Rayleigh distribution; $P(x) = \frac{2x}{\bar{x}^2}\exp(-\frac{x^2}{\bar{x}^2})$, we have $\phi(r) = \exp(-\frac{r^2}{\bar{x}^2})$. Taking a new variable $\theta = \frac{r^2}{\bar{x}^2}$, the differentiation of (C-3) is reduced to

$$\left(\frac{1}{2\theta}-1\right)+\frac{(N-1)\exp(-\theta)}{1-\exp(-\theta)}. \qquad (C\text{-}4)$$

An approximation leads to the solution

$$\theta = \ln N + 0(\frac{1}{\theta}). \qquad (C\text{-}5)$$

This gives an estimate of the maximum value of x in terms of the *rms* value of $\bar{x}$ and the number N.

Appendix D: Static Estimate of Seismic Energy

Suppose that the stress change is given by $\sigma_{ij}(\vec{\xi})$ and that the displacement associated with the stress change is represented by $u_i(\vec{\xi},t)$ at a point $\vec{\xi}$ and a time t. The total work is defined as (Rudnicki and Freund, 1981)

$$W_0 = -\int_{-\infty}^{\infty} dt \int_S dS\sigma_{ij}\gamma_j\dot{u}_i, \qquad (D\text{-}1)$$

where the dot stands for time derivative, S is the source area, and γ_j is the unit normal to S.

If an earthquake source is considered as shear faulting with displacement in only the ξ_1 direction, $u_i(\vec{\xi},t) = u(\vec{\xi},t)\delta_{i1}$, where δ_{ij} is Kronecker's delta and the stress change is on only the $\xi_1\xi_2$-plane. For a local stress drop $\Delta\sigma(\vec{\xi})$ with opposite sign to σ_{ij} in (D-1), and an associated dislocation $D(\xi,t)$ on a heterogeneous fault plane, we have

$$\Delta\sigma(\vec{\xi}) = \Delta\sigma_0 + \delta\sigma(\vec{\xi}), \qquad (D\text{-}2)$$
$$\dot{D}(\vec{\xi},t) = \bar{a} + \delta\dot{D}(\vec{\xi},t), \qquad (D\text{-}3)$$

where $\Delta\sigma_0$ is the average stress drop and $\bar{a}$ is the average dislocation velocity on the fault. Variables $\delta\sigma$ and $\delta\dot{D}$ represent the fluctuation of stress drop and dislocation velocity on random fault patches, and have zero mean. The work done by this heterogeneous faulting is

$$\begin{aligned} W_h &= \frac{1}{2}\int_{-\infty}^{\infty} dt \int_S dS \Delta\sigma(\vec{\xi})\dot{D}(\vec{\xi},t) \\ &= \frac{1}{2}\int_{-\infty}^{\infty} dt \int_S dS \big\{\Delta\sigma_0\bar{a} + \delta\sigma(\vec{\xi})\delta\dot{D}(\vec{\xi},t)\big\}. \end{aligned} \qquad (D\text{-}4)$$

The second term in the right hand side of the above equation is bounded by Cauchy-Schwarz's inequality, and we have

$$\delta W \leq \frac{1}{2}\Big\{\int_S dS \int dt\ \delta\sigma^2(\vec{\xi})\Big\}^{1/2} \Big\{\int_S dS \int dt\ \delta\dot{D}^2(\vec{\xi},t)\Big\}^{1/2}. \qquad (D\text{-}5)$$

From the definition of variance stress drop in (2-16) and from the definition of variance dislocation velocity in (2-5), (D-5) is rewritten

$$\delta W \leq \frac{1}{2}\Big\{\frac{<\Delta\sigma^2>^2}{2\mu^2} L^2W^4\ \lambda T_0\Big\}^{1/2}. \qquad (D\text{-}6)$$

Considering seismic moment in terms of stress drop in (2-15), (D-6) reduces to

$$\delta W \leq \frac{\Delta\sigma_0}{2\mu} M_o\Big\{\frac{<\Delta\sigma^2>}{\Delta{\sigma_0}^2}\Big(\frac{\lambda T_0}{2}\Big)^{1/2}\Big\}. \qquad (D\text{-}7)$$

Appendix E: Time-domain Amplitude and Spectral Amplitude

Consider a random variable $x(t)$. If ω_s is a characteristic frequency, the filtered random variable $x(t;\omega_s)$ is

$$\begin{aligned} x(t;\omega_s) &= \int_{-\infty}^{\infty} x(t-\tau)\exp(i\omega_s\tau)d\tau \\ &= \frac{1}{2\pi}\int_{-\infty}^{\infty} X(\omega)\delta(\omega-\omega_s)\exp(i\omega t)d\omega, \end{aligned} \qquad (E\text{-}1)$$

where $X(\omega)$ is the Fourier transform of $x(t)$

$$X(\omega) = \int_{-\infty}^{\infty} x(t) \exp(-i\omega t) dt, \qquad (E\text{- }2)$$

and δ is Dirac delta function. (E-1) is directly derived from the convolution theorem for Fourier transform.

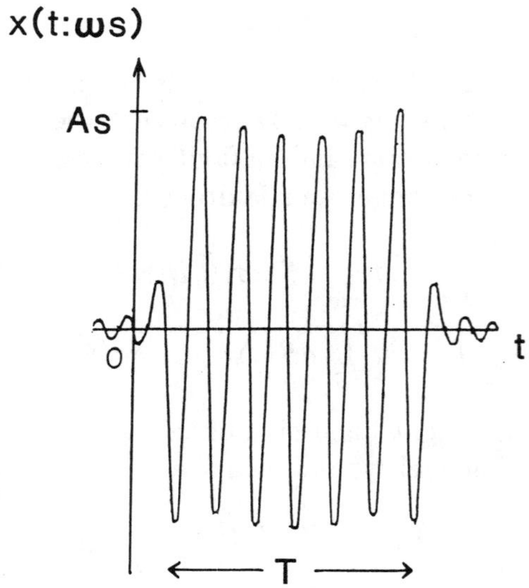

Figure E- 1. An example of filtered signal $x(t;\omega_s)$ with continuous phase. Signal duration is T.

Suppose that $x(t;\omega_s)$ represents coherent waves with a signal duration T. Figure E-1 illustrates such a waveform. Then the time domain amplitude A_s of the waves provides an estimate of the spectrum

$$\begin{aligned}\left|\int_{-T/2}^{T/2} x(t;\omega_S) \exp(-i\omega t) dt\right| &= \left|\int_{-T/2}^{T/2} A_s \exp(i[\omega_s t + \phi]) \exp(-i\omega t) dt\right| \\ &\simeq A_s T, \qquad (E\text{- }3)\end{aligned}$$

where ϕ is the phase.

From (E-1) the spectrum of the filtered variable is

$$\begin{aligned}&\int_{-T/2}^{T/2} x(t;\omega_s) \exp(-i\omega t) dt \\ &= \int_{-T/2}^{T/2} dt \exp(-i\omega t) \frac{1}{2\pi} \int_{-\infty}^{\infty} d\omega' X(\omega') \delta(\omega' - \omega_s) \exp(i\omega' t) \\ &= \frac{1}{2\pi} \int_{-\infty}^{\infty} d\omega' X(\omega') \delta(\omega' - \omega_s) \frac{2\sin(\omega - \omega')T/2}{\omega - \omega'}. \qquad (E\text{- }4)\end{aligned}$$

If the signal duration T is very small, the fraction in the right hand side of (E-4) approaches T. Equating (E-3) and (E-4) yields a relation for the time domain amplitude

$$A_s = \frac{|X(\omega_s)|}{2\pi}. \qquad (E\text{-}5)$$

For large T, the fraction in (E-4) is approximated as

$$\begin{aligned} \lim_{T\to\infty} \frac{\sin xT}{\pi x} &= \frac{1}{2\pi}\int_{-\infty}^{\infty} \exp(ixt)dt \\ &= \delta(x). \end{aligned} \qquad (E\text{-}6)$$

Therefore the time domain amplitude is related to the spectrum by

$$A_s = \frac{|X(\omega_s)|}{T}. \qquad (E\text{-}7)$$

Note that the right hand side of (E-7) has a dimension of amplitude. (E-7) may not be mathematically obvious. However the formal calculus of hyperfunctions leads to (E-7). This can be proved as follows: δ-function can be defined as the extreme of a Gauss error function

$$\delta(\omega - \omega_s) = \lim_{\Delta\to 0} \frac{1}{\sqrt{2\pi\Delta^2}} \exp\Big(-\frac{[\omega-\omega_s]^2}{2\Delta^2}\Big). \qquad (E\text{-}8)$$

Substitute (E-8) into (E-4), and perform the integration by the steepest descent method. Then by taking the extreme of $\Delta \to 0$, (E-7) is obtained.

Appendix F: Rms Amplitude of Random-phase Seismic Waves

Suppose that $x(t;\omega_s)$ is characterized by a filtered signal with random phase. Examples of such wave trains can be found in Fig. F-1. Autocovariance of the filtered variable can be expressed from (D-1)

$$\begin{aligned} &\int_{-\infty}^{\infty} x(t+\tau;\omega_s)x(t;\omega_s)^* dt \\ &= \int_{-\infty}^{\infty} dt\Big\{\frac{1}{2\pi}\int_{-\infty}^{\infty} X(\omega')\delta(\omega'-\omega_s)\exp(i\omega'\{t+\tau\})d\omega'\Big\} \\ &\quad \times\Big\{\frac{1}{2\pi}\int_{-\infty}^{\infty} X(\omega)\delta(\omega-\omega_s)\exp(i\omega t)d\omega\Big\}^* \\ &= \frac{1}{2\pi}\int_{-\infty}^{\infty} X(\omega)X(\omega)^*\delta(\omega-\omega_s)\exp(i\omega\tau)d\omega, \end{aligned} \qquad (F\text{-}1)$$

where $*$ implies the complex conjugate. Formal calculus can be employed similarly to Appendix D to derive (F-1).

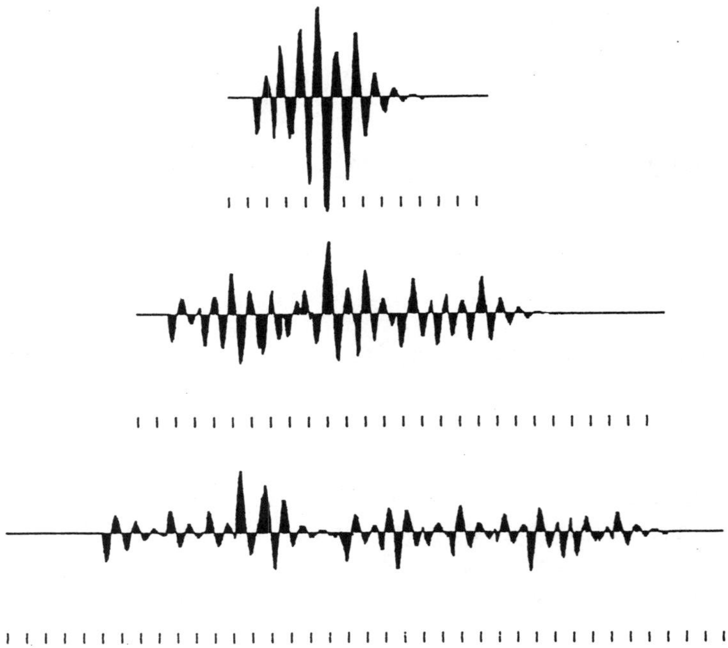

Figure F- 1. Examples of filtered signals with random phases. A component wavelet is characterized by a period of two tick marks. Many of the wavelets are superposed in a random manner. Upper and lower traces include the same number of wavelets, with wavelets more heavily overlapped in the upper.

When τ tends to zero, (F-1) gives Parseval's equality for the filtered variable

$$\begin{aligned}\int_{-\infty}^{\infty} |x(t;\omega_s)|^2 \, dt &= \frac{1}{2\pi}\int_{-\infty}^{\infty} X(\omega)X(\omega)^*\delta(\omega-\omega_s)d\omega \\ &= \frac{1}{\pi}|X(\omega_s)|^2. \qquad (F\text{-}\ 2)\end{aligned}$$

Appendix G: Steepest Descent Approximation

The approximation employed in this study has been developed by Hara et al. (1992). Here we modify the original derivation for the present application. Consider a summation of

$$N(\tau) = \sum_{i=0}^{M} \alpha^i \exp(-\beta^i\, \tau) \qquad (0 \le \tau), \qquad (G\text{-}\ 1)$$

where α and β are positive constants smaller than unity and M is sufficiently large number. In the case $1 \ll \tau$, (G-1) is approximately written

$$N(\tau) \simeq \int_0^\infty \exp(i \ln \alpha - \beta^i \tau) di. \qquad (G\text{-}2)$$

By taking Taylor series approximation of the argument of the exponential function, the integral in (G-2) can be evaluated by the steepest descent path through i_0, where

$$\ln \alpha - \tau \ln \beta \, \beta^{i_0} = 0. \qquad (G\text{-}3)$$

This is justified because the higher order terms in the Taylor series converge more rapidly than the term derived from the steepest descent. Then

$$\begin{aligned} N(\tau) &\simeq \exp(i_0 \ln \alpha - \beta^{i_0} \tau) \int_0^\infty \exp(-\frac{\ln \alpha \ln \beta}{2} [i - i_0]^2) di \\ &\simeq \sqrt{\frac{\pi}{2 \ln \alpha \ln \beta}} \exp(\xi \ln(\frac{\xi}{\tau}) - \xi), \end{aligned} \qquad (G\text{-}4)$$

where ξ is

$$\xi = \frac{\ln \alpha}{\ln \beta}. \qquad (G\text{-}5)$$

REFERENCES

Chapter 1

Aki, K., and P.G. Richards, 1980, *Quantitative seismology Theory and Methods*, Freeman & Company, San Francisco, 63-121.

Jost, M.L., and R.B. Herrmann, 1989, A student's guide to and review of moment tensor, Seismological Research Letters, **60**, 37-57.

Kikuchi M., 1987, Fault models: Source processes (in Japanese), in *The encyclopedia of earthquakes*, T. Utsu ed., Asakura Shoten, Tokyo, 226-242.

Lay, T., and T.C. Wallace, 1995, *Modern global seismology*, Academic Press, San Diego, 310-433.

Sato, T., and T. Hirasawa, 1973, Body wave spectra from propagating shear cracks, Journal of Physics of the Earth, **21**, 415-431.

Savage, J.C., 1972, Relation of corner frequency to fault dimensions, Journal of Geophysical Research, **77**, 3788-3795.

Shimizu, H., S. Ueki and J. Koyama, 1987, A tensile-shear crack model for the mechanism of volcanic earthquakes, Tectonophysics, **144**, 287-300.

Chapter 2

Abe, K., 1975, Reliable estimation of the seismic moment of large earthquakes, Journal of Physics of the Earth, **23**, 381-390.

Aki, K., 1967, Scaling law of seismic spectrum, Journal of Geophysical Research, **72**, 1217-1231.

Kanamori, H., and D.L. Anderson, 1975, Theoretical basis of some empirical relations in seismology, Bulletin of the Seismological Society of America, **65**, 1073-1095.

Koyama, J., 1985, Earthquake source time function from coherent and incoherent rupture, Tectonophysics, **118**, 227-242.

Koyama, J., 1994, General description of the complex faulting process and some empirical relations in seismology, Journal of Physics of the Earth, **42**, 103-148.

Koyama, J., and S.H. Zheng, 1985, Excitation of short-period body waves by great earthquakes, Physics of the Earth and Planetary Interiors, **37**, 108-123.

Lay, T., H. Kanamori and L. Ruff, 1982, The asperity model and the nature of large subduction zone earthquakes, Earthquake Prediction Research, **1**, 3-71.

Takemura, M., and J. Koyama, 1983, A scaling model for low-frequency earthquakes - Relation of source spectra between tsunami earthquakes and small low-frequency earthquakes (in Japanese), Zisin, Ser.2, **36**, 323-333.

Chapter 3

Aki, K., 1989, Research notes of the National Research Center for Disaster Prevention, No.80, 1-130.

Boatwright, J., and D.M. Boore, 1982, Analysis of the ground acceleration radiated by the 1980 Livermore Valley earthquake for directivity and dynamic source characteristics, Bulletin of the Seismological Society of America, **72**, 1843-1865.

Chen, P., and O.W. Nuttli, 1984, Estimations of magnitudes and short-period wave attenuation of Chinese earthquakes from modified Mercalli intensity data, Bulletin of the Seismological Society of America, **74**, 957-968.

Gumbel, E.J., 1958, *Statistics of extremes*, Columbia University Press, New York, 1-254.

Hanks, T.C., 1982, f_{max}, Bulletin of the Seismological Society of America, **72**, 1867-1879.

Jiang, F., 1978, *The Haicheng Earthquake* (in Chinese), Seismological Press, Beijing, 1-90.

Kawasumi, H., and Y. Sato, 1968, Intensity of Niigata earthquake as determined from questionnaires, *General report of the Niigata earthquake, 1964*, Tokyo Electronics Engineering College, 175-179.

Kinoshita, S., 1988, The recent topics pertinent to f_{max} (in Japanese), Zisin, Ser.2, **44**, 301-314.

Koyama, J., and Y. Izutani, 1990, Seismic excitation and directivity of short-period body waves from a stochastic fault model, Tectonophysics, **175**, 67-79.

Koyama, J., and S.H. Zheng, 1991, Seismic intensity distribution of shallow earthquakes due to rupture velocities and faulting modes, Acta Seismologica Sinica, **13**, 190-201.

Mei, S.R., 1982, *The Tangshan earthquake of 1976* (in Chinese), Seismological Press, Beijing, 1-459.

Ohta, Y., H. Kagami and S. Okada, 1987, Seismic intensity and its application to engineering: A study of Japan, In *Strong ground motion seismology*, Erdik, M.O., and M.N. Toksöz eds., Reidel Publisher's Group, Dordrecht, 369-384.

Papageorgiou, A.S., and K. Aki, 1983, A specific barrier model for quantitative description of inhomogeneous faulting and the prediction of strong ground motion. Part 1. Description of the model, Bulletin of the Seismological Society of America, **73**, 693-722.

Rice, J.R., 1980, The mechanics of earthquake rupture, Proceeding of international school of physics, *Enrico Fermi* Italian Physical Society, LXXVIII, Boschi, E., and A.M. Dziewonski eds., North-Holland Publishing Company, Amsterdam, 555-649.

State Seismological Bureau, 1979, *The Isoseismal Maps of Chinese Earthquakes* (in Chinese), Seismological Press, Beijing, 1-107.

Takemura, M., M. Motosaka and H. Yamanaka, 1995, Strong motion seismology in Japan, Journal of Physics of the Earth, **43**, 211-257.

Chapter 4

Abe, K., 1981, Magnitudes of large shallow earthquakes from 1904-1980, Physics of the Earth and Planetary Interiors, **27**, 72-92.

Boatwright, J., and G.L. Choy, 1986, Teleseismic estimates of the energy radiated by shallow earthquakes, Journal of Geophysical Research, **91**, 2095-2112.

Haskell, N.A., 1964, Total energy and energy spectral density of elastic wave radiation from propagating faults, Bulletin of the Seismological Society of America, **54**, 1811-1841.

Houston, H., and H. Kanamori, 1986, Source spectra of great earthquakes: Teleseismic constraints on rupture process and strong motion, Bulletin of the Seismological Society of America, **76**, 19-42.

Kanamori, H., 1977, The energy release in great earthquakes, Journal of Geophysical research, **82**, 2981-2987.

Kikuchi, M., and Y. Fukao, 1988, Seismic wave energy inferred from long-period body wave inversion, Bulletin of Seismological Society of America, **78**, 1707-1724.

Chapter 5

Ekström, G., and A.M. Dziewonski, 1988, Evidence of bias in estimations of earthquake size, Nature, **332**, 319-323.

Gutenberg, B., 1945, Amplitudes of surface waves and magnitudes of shallow earthquakes, Bulletin of the Seismological Society of America, **35**, 3-12.

Koyama, J., and N. Shimada, 1985, Physical basis of earthquake magnitudes: An extreme value of seismic amplitudes from incoherent fracture of random fault patches, Physics of the Earth and Planetary Interiors, **40**, 301-308.

Purcaru, G., and H. Berckhemer, 1982, Quantitative relations of seismic source parameters and a classification of earthquakes, Tectonophysics, **84**, 57-128.

Chapter 6

Aki, K., 1972, Scaling law of earthquake source time-function, Geophysical Journal of the Royal Astronomical Society, **31**, 3-25.

Anderson, J.G., P. Bodin, J.N. Brune, J. Prince, S.K. Singh, R. Quaas and M. Onate, 1986, Strong ground motion from the Mechoacan, Mexico, earthquake, Science, **233**, 1043-1049.

Brune, J.N., 1970, Tectonic stress and the spectra of seismic shear waves from earthquakes, Journal of Geophysical Research, **75**, 4997-5009.

Vassiliou, M.S., and H. Kanamori, 1982, The energy release in earthquakes, Bulletin of the Seismological Society of America, **72**, 371-387.

Chapter 7

Amit, D.J., 1989, *Modeling Brain Function*, Cambridge University Press, New York, 58-96.

Goto, K., S. Mori and K. Fukui, 1991, Volcanic tremor accompanying the 1989 submarine eruption off the east coast of Izu peninsula, Japan, Journal of Physics of the Earth, **39**, 47-63.

Hida, T., 1980, *Brownian motion*, Springer, New York.

Koyama, J., and H. Hara, 1992, Scaled Langevin equation to describe $1/f^{\alpha}$ spectrum, Physical Review **A 46**, 1844-1849.

Koyama, J., and H. Hara, 1993, Fractional Brownian motion described by scaled Langevin equation, Chaos, Solitons & Fractals, **3**, 467-480.

Mandelbrot, B.B., and J.W. Van Ness, 1968, Fractional Brownian motions, fractional Noises and applications, SIAM Review, **10**, 422-437.

Montroll, E.W., and M.F. Shlesinger, 1984, On the wonderful world of random walks, In *Nonequilibrium Phenomena II From Stochastics to Hydrodynamics*, J.L. Lebowiz and E.W. Montroll eds., Elsevier, Amsterdam, 1-121.

Musha, T., S. Sato and M. Yamamoto, 1991, *Proceedings of the International Conference on Noise in Physical Systems and 1/f Fluctuations*, Ohmusha, Tokyo, 1-752.

Chapter 8

Bak, P., and C. Tang, 1989, Earthquakes as a self-organized critical phenomenon, Journal of Geophysical Research, **94**, 15635-15637.

Carlson, J.M., 1991, Time intervals between characteristic earthquakes and correlations with smaller events: An analysis based on a mechanical model of a fault, Journal of Geophysical Research, **96**, 4255-4267.

Cheng, Y.T. and L. Knopoff, 1987, Simulation of earthquake sequences, Geophysical Journal of Royal Astronomical Society, **91**, 693-709.

Koyama, J., and Y. Nakamura, 1980, Focal mechanism of deep moonquakes, Proceedings of the Eleventh Lunar and Planetary Science Conference, Pergamon Press, New York, 1855-1865.

Montroll, E.W., and M.F. Shlesinger, 1983, Maximum entropy formalism, fractals, scaling phenomena, and $1/f$ noise: A tale of tails, Journal of Statistical Physics, **32**, 209-230.

Nakamura, Y., 1978, A_1 moonquakes: Source distribution and mechanism, Proceedings of the Ninth Lunar and Planetary Science Conference, Pergamon Press, New York, 3589-3607.

Ogata, Y., and K. Abe, 1991, Some statistical features of the long-term variation of the global and regional seismic activity, International Statistical Review, **59**, 139-161.

Ohtsuka, M., 1972, A chain-reaction type source model as a tool to interpret the magnitude-frequency relation of earthquakes, Journal of Physics of the Earth, **20**, 35-45.

Ouchi, T., 1993, Population dynamics of earthquakes and mathematical modeling, PAGEOPH, **140**, 15-28.

Utsu, T., Y. Ogata and R.S. Matsu'ura, 1995, The centenary of the Omori formula for a decay law of aftershock activity, Journal of Physics of the Earth, **43**, 1-33.

Chapter 9

Hara, H., and J. Koyama, 1992, Activation processes of complex system and correlation function (in Japanese), Proceeding of the Institute of Statistical Mathematics, **40**, 217-226.

Hara, H., T. Obata and Y. Tamura, 1996, Inverse problem on Riemann-Liouville Integral (in Japanese), Proceeding of the Institute of Statistical Mathematics, in press.

Matsuo, T., R. Okeda, M. Takahashi and M. Funata, 1990, Characterization of bifurcating structures of blood vessels using fractal dimensions, Forma, **5**, 19-27.

Nakamura, Y., 1980, Shallow moonquakes: How they compare with earthquakes, Proceedings of the Eleventh Lunar Planetary Science Conference, Pergamon Press, New York, 1847-1853.

Pacheco, J.F., C.H. Scholz and L.R. Sykes, 1992, Changes in frequency-size relationship from small to large earthquakes, Nature, **355**, 71-73.

Appendix A

Bendat, J.S., 1958, *Principles and applications of random noise theory*, John Wiley & Sons, New York, 1-218.

Appendix D

Rudnicki, J.W., and L.B. Freund, 1981, On energy radiation from seismic sources, Bulletin of the Seismological Society of America, **71**, 583-595.

Appendix G

Hara, H., O.K. Chung and J. Koyama, 1992, Dynamical activation processes described by generalized random walks, Physical Review B, **46**, 838-845.

LIST OF SYMBOLS

a_d, a_s	scaling parameters
$\bar{a}$	average dislocation velocity
$\vec{a}$	eigenvector
a_{max}, A_{max}	maximum amplitude in time domain
a_{rms}, A_{rms}	root-mean-square amplitude
b	b-value of Gutenberg and Richter' relation
b_d, b_s	scaling parameters
b_{rms}, b_{max}	rms and maximum amplitudes of bilateral faulting
$B_c(\omega)$	displacement source spectrum
c	P or S-wave velocity
c_0	correction for source amplitude
$C(\xi;\tau)$	autocovariance of dislocation function
$C(\tau)$, $C_j(\tau)$	autocovariance of local states
d	power-coefficient of time interval distribution
$\bar{d}$	characteristic fault-patch size
D	dislocation time function
D_0	average dislocation
$\vec{D}$	vector of dislocation function
$\dot{D}$	dislocation velocity function
$\vec{e}(e_1, e_2, e_3)$	slip direction
$\vec{e}_r$, $\vec{e}_\theta$, $\vec{e}_\phi$	unit vectors at observation
E_S, E_S^P, E_S^S	total seismic, P- and S-wave energies
E_T	entropy
$E[\ ]$	expectation
f	frequency
f_{max}	characteristic cut-off frequency in Hz
$f(N)$	function from statistical theory of extremes
$F(t)$	time function of a single force
$F(\omega)$	seismic directivity function
$< f_P >$, $< f_S >$	corner frequency of P- and S-waves
$g(\chi)$	function due to rupture velocity
G	geometrical spreading
G_h, $G_f(\omega)$	stochastic, fractal acceleration spectra
$h(t)$, $h_j(t)$	impulse response
H	Hurst exponent
$H(t)$	Heaviside step function
$\boldsymbol{I}$	identity matrix
I_J	Japan Meteorological Agency intensity scale
I_m	model seismic intensity
I_M	modified Mercalli intensity

$\vec{l}\,(l_1, l_2, l_3)$	unit vector of a single force
$L,\ W,\ S$	fault length, width, area size
$m_B,\ m_b$	body-wave magnitude
$m_S,\ m_T$	seismic moment of shear, tensile crack
$\boldsymbol{M}$	moment tensor
M_1	short-period seismic excitation
M_L	local magnitude (Richter scale)
M_{jk}	moment tensor
M_o	seismic moment(scalor)
$M_o(t),\ \dot{M}_o(t)$	seismic moment, moment rate function
$M_S,\ M_s$	surface-wave magnitude
$\vec{n}(\nu_1, \nu_2, \nu_3)$	fault normal
$n(t)$	number of aftershocks
$n_0(t), n_i(t)$	random force
$N,\ N_a,\ N_b$	number of peaks and troughs
p	ratio of W/L or Omori-Utsu power coefficient
P_a	total power of acceleration
P_t	total power of acceleration spectrum
$P_0,\ P_j$	probability density
q	ratio of T_0 to $W/\bar{v}$
r_0	distance to observation from origin
$R_j(t)$	complex random noise
$R_{\theta\phi}$	radiation pattern coefficient
S_0	high-frequency source spectrum
t^*	travel time divided by quality factor
T_0	rise time for dislocation
T_d	time duration of acceleration wave train
$\vec{u}$	displacement vector
$u_1,\ u_2,\ u_3$	displacement components(Cartesian)
$u_r,\ u_\theta,\ u_\phi$	displacement components(spherical)
$\bar{v}$	average rupture velocity
$W_0,\ W_h$	energy available for seismic waves
W_t	total work done by an earthquake
$\vec{x}(x_1, x_2, x_3)$	observation vector
$X(t),\ X_j(t)$	state of the system, local states
$Z_j(t)$	complex-valued local states
$\alpha,\ \beta$	P-, S-wave velocity
γ	Euler constant
$\bar{\gamma}$	characteristic constant
$\vec{\gamma}(\gamma_1,\ \gamma_2,\ \gamma_3)$	unit vector to observation
Γ	Gamma function
δ	dip angle

$\delta(t)$	Dirac delta function
δ_{ij}	Kronecker's delta
Δ	epicentral distance
$\Delta\sigma_0$	average stress drop
$\Delta\sigma_0(\vec{\xi})$	local stress drop
$< \Delta\sigma^2 >$	variance stress drop
$\epsilon,\ \varepsilon$	constant for M_S definitions
ε_b	constant for m_b definition
$\zeta_1,\ \zeta_2$	characteristic angular frequencies of fault patches
$\eta_0,\ \eta_h$	seismic efficiency
η_M	maximum power coefficient
$\theta,\ \phi$	polar angle, azimuth angle
κ	fractal dimension
$\tilde{\kappa}$	complex fractal dimension
λ	patch corner frequency or slip angle
$\mu,\ \rho$	rigidity, density
μ_0	fracture probability
$\nu_1,\ \nu_2,\ \nu_3$	components of fault normal
σ^2	variance dislocation velocity
$\sigma_1,\ \sigma_2,\ \sigma_f$	initial, final and frictional stresses
χ	ratio of $\bar{v}/\beta$
ω	angular frequency
ω_b	characteristic angular frequency for body-waves
$\omega_c,\ \omega_\beta$	angular corner frequency, that for S-waves
ω_m	cut-off angular frequency $2\pi f_{max}$
ω_s	angular frequency for surface-waves, $2\pi/20$
$\Omega(\omega)$	earthquake source spectrum
ξ_d	fractal dimension for earthquake activity
ξ_s	scaling dimension for earthquake statistics
$\xi_1,\ \xi_2,\ \xi_3$	Cartesian coordinate at source
$< >$	mean value

INDEX

MODERN APPROACHES IN GEOPHYSICS

1. E.I. Galperin: *Vertical Seismic Profiling and Its Exploration Potential.* 1985 ISBN 90-277-1450-9
2. E.I. Galperin, I.L. Nersesov and R.M. Galperina: *Borehole Seismology.* 1986 ISBN 90-277-1967-5
3. Jean-Pierre Cordier: *Velocities in Reflection Seismology.* 1985 ISBN 90-277-2024-X
4. Gregg Parkes and Les Hatton: *The Marine Seismic Source.* 1986 ISBN 90-277-2228-5
5. Guust Nolet (ed.): *Seismic Tomography.* 1987 Hb: ISBN 90-277-2521-7; Pb: ISBN 90-277-2583-7
6. N.J. Vlaar, G. Nolet, M.J.R. Wortel and S.A.P.L. Cloetingh (eds.): *Mathematical Geophysics.* 1988 ISBN 90-277-2620-5
7. J. Bonnin, M. Cara, A. Cisternas and R. Fantechi (eds.): *Seismic Hazard in Mediterranean Regions.* 1988 ISBN 90-277-2779-1
8. Paul L. Stoffa (ed.): *Tau-p: A Plane Wave Approach to the Analysis of Seismic Data.* 1989 ISBN 0-7923-0038-6
9. V.I. Keilis-Borok (ed.): *Seismic Surface Waves in a Laterally Inhomogeneous Earth.* 1989 ISBN 0-7923-0044-0
10. V. Babuska and M. Cara: *Seismic Anisotropy in the Earth.* 1991 ISBN 0-7923-1321-6
11. A.I. Shemenda: *Subduction.* Insights from Physical Modeling. 1994 ISBN 0-7923-3042-0
12. O. Diachok, A. Caiti, P. Gerstoft and H. Schmidt (eds.): *Full Field Inversion Methods in Ocean and Seismo-Acoustics.* 1995 ISBN 0-7923-3459-0
13. M. Kelbert and I. Sazonov: *Pulses and Other Wave Processes in Fluids.* An Asymptotical Approach to Initial Problems. 1996 ISBN 0-7923-3928-2
14. B.E. Khesin, V.V. Alexeyev and L.V. Eppelbaum: *Interpretation of Geophysical Fields in Complicated Environments.* 1996 ISBN 0-7923-3964-9
15. F. Scherbaum: *Of Poles and Zeros.* Fundamentals of Digital Seismology. 1996 Hb: ISBN 0-7923-4012-4; Pb: ISBN 0-7923-4013-2
16. J. Koyama: *The Complex Faulting Process of Earthquakes.* 1997 ISBN 0-7923-4499-5

KLUWER ACADEMIC PUBLISHERS – DORDRECHT / BOSTON / LONDON